Springer Theses

Recognizing Outstanding Ph.D. Research

Aims and Scope

The series "Springer Theses" brings together a selection of the very best Ph.D. theses from around the world and across the physical sciences. Nominated and endorsed by two recognized specialists, each published volume has been selected for its scientific excellence and the high impact of its contents for the pertinent field of research. For greater accessibility to non-specialists, the published versions include an extended introduction, as well as a foreword by the student's supervisor explaining the special relevance of the work for the field. As a whole, the series will provide a valuable resource both for newcomers to the research fields described, and for other scientists seeking detailed background information on special questions. Finally, it provides an accredited documentation of the valuable contributions made by today's younger generation of scientists.

Theses may be nominated for publication in this series by heads of department at internationally leading universities or institutes and should fulfill all of the following criteria

- They must be written in good English.
- The topic should fall within the confines of Chemistry, Physics, Earth Sciences, Engineering and related interdisciplinary fields such as Materials, Nanoscience, Chemical Engineering, Complex Systems and Biophysics.
- The work reported in the thesis must represent a significant scientific advance.
- If the thesis includes previously published material, permission to reproduce this must be gained from the respective copyright holder (a maximum 30% of the thesis should be a verbatim reproduction from the author's previous publications).
- They must have been examined and passed during the 12 months prior to nomination.
- Each thesis should include a foreword by the supervisor outlining the significance of its content.
- The theses should have a clearly defined structure including an introduction accessible to new PhD students and scientists not expert in the relevant field.

Indexed by zbMATH.

More information about this series at https://link.springer.com/bookseries/8790

Yasuomi Yamazaki

Development of Synthetic Methods for Novel Photofunctional Multinuclear Complexes

Simple Synthetic Methods for Multinuclear Complexes Using Various C-C Coupling Reactions

Doctoral Thesis accepted by
Tokyo Institute of Technology, Tokyo, Japan

Springer

Author
Dr. Yasuomi Yamazaki
Department of Materials and Life Sciences
Faculty of Science and Technology
Seikei University
Tokyo, Japan

Supervisor
Prof. Osamu Ishitani
Department of Chemistry
School of Science
Tokyo Institute of Technology
Tokyo, Japan

ISSN 2190-5053 ISSN 2190-5061 (electronic)
Springer Theses
ISBN 978-981-16-7147-0 ISBN 978-981-16-7148-7 (eBook)
https://doi.org/10.1007/978-981-16-7148-7

This Springer imprint is published by the registered company Springer Nature Singapore Pte Ltd.
The registered company address is: 152 Beach Road, #21-01/04 Gateway East, Singapore 189721, Singapore

Supervisor's Foreword

Global warming and the depletion of fossil resources are serious threats to the future of human beings. The development of useful photofunctional compounds for solar energy conversion is an urgent priority for achieving a sustainable society. One of the useful strategies should be an integration of different metal complexes with different photofunctions, which can induce various processes such as excitation energy and electron transfer, and catalytic reactions. From this viewpoint, Dr. Yamazaki's Ph.D. work has a strong impact on science and technology, especially for improving the conversion efficiency of solar energy to chemical energy.

The thesis written by Dr. Yasuomi Yamazaki addresses new synthetic methods for a novel class of various photofunctional multinuclear complexes. He applied the various C-C coupling methods, which have been originally developed for the synthesis of organic compounds, to the synthesis of photofunctional multinuclear metal complexes. By using these synthetic methods, he successfully made various types of multinuclear complexes with high photofunctions, such as light-harvesting properties and photocatalyses. For instance, he developed various photocatalysts consisting of photosensitizer and catalyst units, of which photocatalytic abilities for CO_2 reduction are much higher compared to the corresponding mixed system of mononuclear complexes due to rapid intramolecular electron transfer between the photosensitizer and catalyst units. Light-harvesting multinuclear complexes are the other important examples, where rapid intramolecular energy transfer accumulates excitation energy into one unit.

The main contents of the Ph.D. thesis of Dr. Yasuomi Yamazaki have already been published in 7 high-level international journals such as *Chemical Science* and *Inorganic Chemistry*.

Tokyo, Japan
March 2021

Osamu Ishitani

List of Published Articles

Parts of this thesis have been published in the following journal articles:

(1) Y. Yamazaki, T. Morimoto, O. Ishitani, "Synthesis of Novel Photofunctional Multinuclear Complexes Using a Coupling Reaction", *Dalton Trans.*, **2015**, *44*, 11626–11635.

(2) Y. Yamazaki, A. Umemoto, O. Ishitani, "Photochemical Hydrogenation of π-Conjugated Bridging Ligands on Photofunctional Multinuclear Complexes", *Inorg. Chem.*, **2016**, *55*, 11110–11124.

(3) Y. Yamazaki, O. Ishitani, "Selectivity control between Mizoroki-Heck and homo-coupling reactions for synthesising multinuclear metal complexes: unique addition effects of tertiary phosphines and O_2", *Dalton Trans.*, **2017**, *46*, 4816–4823.

(4) Y. Yamazaki, O. Ishitani, "Synthesis of Os(II)–Re(I)–Ru(II) hetero-trinuclear complexes and their photophysical properties and photocatalytic abilities", *Chem. Sci.*, **2018**, *9*, 1031–1041.

(5) Y. Yamazaki, J. Rohacova, H. Ohtsu, M. Kawano, O. Ishitani, "Synthesis of Re(I) Rings Comprising Different Re(I) Units and Their Light-Harvesting Abilities", *Inorg. Chem.*, **2018**, *57*, 15158–15171.

(6) Y. Yamazaki, O. Ishitani, "Synthesis of an Emissive Spectacle-Shaped Hexanuclear Rhenium(I) Complex", *Inorg. Chem.*, **2019**, *58*, 12905–12910.

(7) Y. Yamazaki, J. Rohacova, K. Koike, O. Ishitani, "Synthesis and Light-Harvesting Functions of Ring-Shaped Re(I) Trinuclear Complexes Connected with an Emissive Ru(II) Complex", *JACS Au*, **2021**, *1*, 294–307.

Acknowledgements

First and foremost, I would like to express my sincere thanks to my supervisor, Prof. Osamu Ishitani, for his continuous discussion, suggestions, and supports. I have respect for his great effort in research and education. I will build on the variable experiences in the Ishitani laboratory to work as a researcher. I look forward to continuous collaboration work with him.

Besides my supervisor, I wish to thank Prof. Kazuhiko Maeda and Prof. Tatsuki Morimoto for their countless encouragement and insightful comments in the laboratory meetings. I gratefully acknowledge the work of past and present members of the Ishitani laboratory.

I am grateful to Prof. Masaki Kawano and Prof. Hiroyoshi Ohtsu for the X-ray crystal analysis. I also thank Prof. Hiromu Kumagai, Prof. Hiroyuki Takeda, Dr. Jana Rohacova, and Dr. Kazuhide Koike for the collaboration works.

Last, I would also like to express my gratitude to my family for their moral support and warm encouragement.

March 2021

Contents

Chapter 1
General Introduction

Abstract Light-induced electron and/or energy transfer are essential reactions for solar energy conversion, which is an urgent priority to achieve a sustainable society. In addition to the physicochemical research, synthetic chemistry for synthesizing various series of multifarious photofunctional compounds is contributable for understanding and controlling those processes. Hetero-multinuclear metal complexes should be one of the most interesting photofunctional compounds because they often show useful photophysical properties, photochemical behaviors, and photocatalytic abilities due to the efficient intramolecular excitation energy and/or electron transfer and the flexibility in the molecular design. This chapter overviews the backgrounds and analysis of the problems in the conventional synthetic methods for integrating different types of photofunctional metal complexes to form hetero-multinuclear complexes, and then explores new strategies for novel classes of multinuclear complexes.

Keywords Photofunctional metal complexes · Multinuclear complexes · Bridging ligands · Building block approach · Coupling reactions

1.1 Background

Photofunctional multinuclear transition-metal complexes, which are synthesized by linking mononuclear complexes through a bridging ligand or ligands, have attracted great interest because of their potential use for dye-sensitized solar cells [1] and light-harvesting systems [2], owing to their strong absorption abilities over the visible region and/or rapid intramolecular energy hopping. Hetero-multinuclear metal complexes constructed using different functional units have recently been applied as efficient photocatalysts for oxidation and/or reduction of water [3–7], and CO_2 reduction [8–13]. For example, Ru(II)–Re(I) dinuclear complexes, where the Ru(II) and Re(I) units respectively function as a photosensitizer and a catalyst, are effective photocatalysts for CO_2 reduction because of rapid intramolecular electron transfer from the photochemically-reduced Ru unit to the Re(I) unit [10].

One of the most useful synthetic methods of such multinuclear metal complexes is stepwise synthesis via several complexation steps using multidentate bridging ligands

Y. Yamazaki, *Development of Synthetic Methods for Novel Photofunctional Multinuclear Complexes*, Springer Theses, https://doi.org/10.1007/978-981-16-7148-7_1

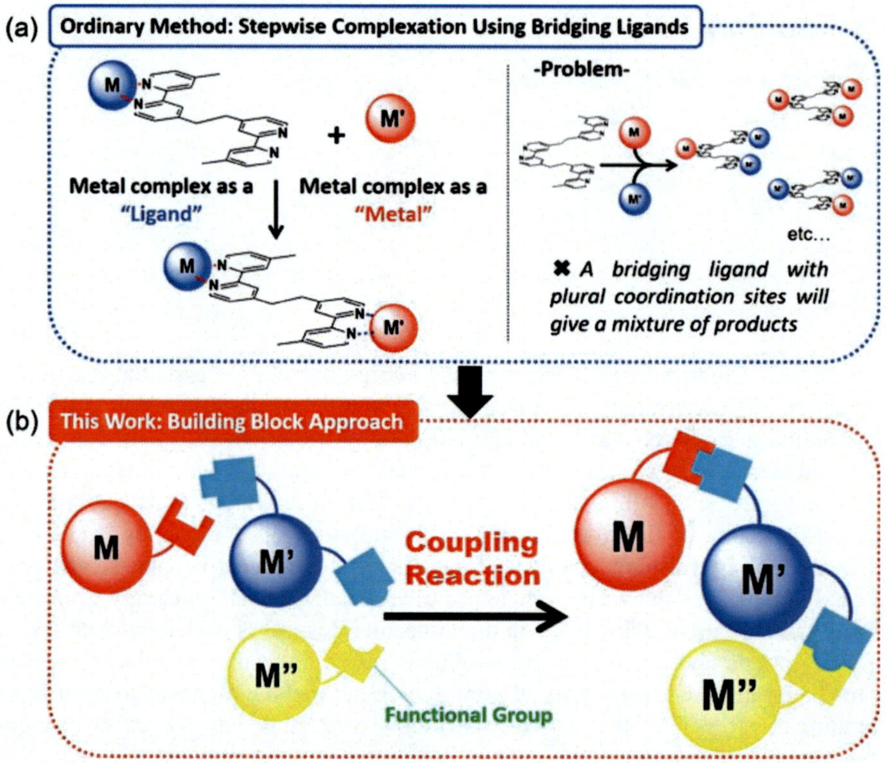

Scheme 1.1 **a** Ordinary and **b** new synthetic methods (*building block approach*) for constructing multinuclear metal complexes

and precursor-complexes [14]. A multidentate ligand, of which one chelating moiety already coordinates to one metal center, is firstly synthesized, and then it reacts with another metal center giving a dinuclear complex (Scheme 1(a)). Various multinuclear complexes have been synthesized by repeating these processes; however, there are some limitations on product selectivity and structure. For example, multidentate ligands with more than two coordination sites might give a mixture of products with different numbers of metal complexes. Especially when the multidentate ligands have different kinds of coordination sites, selective synthesis of one of the diastereoisomers is difficult to be achieved. Therefore, in principle, each bridging ligand can be utilized only for connection between two metal complexes. The number of the bridging ligand of which coordination can be accepted by the precursor complex strongly limits the number of nuclear of the complexes. For example, Ru(II) bis-diimine complexes, which are known as efficient catalysts for CO_2 reduction, could be easily connected with one or two Ru photosensitizer complex(es) via one or two bridging ligand(s) to produce dinuclear or trinuclear photocatalysts [15]. The trinuclear Ru complexes showed higher photocatalytic ability than that of dinuclear complexes because the trinuclear complexes can accumulate two electrons, which are required for reducing

CO_2 to formate, into the catalyst unit more frequently than the dinuclear complexes. While most of the reported photocatalysts with a Re(I) mono-diimine complex as a catalyst unit have possessed only one photosensitizer unit [8, 10, 12, 13, 16–22]. The effect of the number of the photosensitizer units on the photocatalytic ability of photocatalysts with Re catalysts has not been investigated. These synthetic limitations prevent synthesizing and investigating new series of multinuclear complexes.

Another problem is that only complexes which can accept coordination of the bridging ligands can be used as starting materials for this synthetic method. Therefore, most of the multinuclear complexes obtained by using this method cannot be connected with other metal complexes furthermore. For example, ring-shaped multinuclear Re(I) complexes, which show various useful photofunctions, i.e., efficient photosensitizing ability [23, 24] and electron accumulation ability [25], have not yet been connected with other metal complexes even though they show high potentials to function as efficient units for photosensitizers, electron pools or antenna moieties in photofunctional compounds. This is because they are thermally stable enough to suppress ligand substitution reactions with bridging ligands. If useful multinuclear complexes can be easily introduced into photofunctional compounds, their photofunctions will be improved.

In this research, new synthetic methods for connecting metal complexes, the so-called "*building block approach*", were applied for constructing various photofunctional multinuclear complexes (Scheme 1b). Metal complexes with proper functional groups, e.g., vinyl groups, bromo groups, and ethynyl groups, are used as building blocks to connect each other by using coupling reactions. In particular, cross-coupling reactions enable to determine the arrangement of each complex unit in hetero-multinuclear complexes by introducing different functional groups into each metal complex as a building block; therefore, single multinuclear complexes should be selectively obtained as products. In addition, when having plural functional groups, the complex should be connected with several metal complexes in a single step. Various compounds having proper functional groups can be used as starting materials regardless of the structure of their ligands, charges of the compounds, and kinds of metal centers. Therefore, not only mononuclear complexes but also multinuclear complexes are expected to be connected with other metal complexes. By using this approach, various and new photocatalysts and/or photofunctional complexes, such as photocatalysts with several photosensitizer units or emissive complexes with multinuclear complexes as antenna units, will be able to be synthesized.

Some coupling reactions have already been investigated whether they are useful for the connection of photofunctional mononuclear metal complexes, e.g., Sonogashira coupling [26–28], Suzuki–Miyaura coupling [29–31], and homo-coupling [32]. I first tried to use the Mizoroki–Heck reaction, because synthetic procedures for various metal complexes having vinyl or halogen groups in their ligands were already established. In addition to the Mizoroki–Heck reaction, homo-coupling reactions and Sonogashira coupling reactions were also applied for the synthesis of photofunctional multinuclear metal complexes in this thesis.

In most cases using the coupling reactions, each metal complex is connected through an unsaturated bond, e.g., a double or a triple bond and a phenylene group, which often causes extension of π conjugation in the bridging ligand. The introduction of such π conjugation should strongly affect photophysical properties of the synthesized multinuclear complexes: red-shift of absorption and extension of emission lifetime [12, 33], are desirable properties as efficient light-absorbers for the dye-sensitized solar cells and the light-harvesting systems. However, such large π conjugation also causes a positive shift of the redox potential of the multinuclear metal complexes, which often leads to lower photocatalytic ability because of the lower reduction power of the catalyst units. Therefore, in order to use the multinuclear complexes as photocatalysts, the unsaturated bonds in the bridging linker should be reduced into the corresponding saturated ones, which are more suitable bridging linkers for photocatalytic reactions. If the vinylene linkers formed by the Mizoroki–Heck reaction can be converted to saturated carbon chains using a hydrogenation reaction, the multinuclear complexes will also function as efficient photocatalysts. Thus, a combination of the Mizoroki–Heck reaction and hydrogenation will enlarge not only a variety of the multinuclear complexes but also their photofunctions.

From viewpoints as described above, I investigated the following targets:

A. How to integrate various metal complexes to synthesize a new type of photocatalysts and/or photofunctional multinuclear complexes.
B. How to select the binding mode of linkage in the bridging ligands in order to control the photophysical properties and photocatalytic ability of synthesized multinuclear metal complexes.
C. How to use multinuclear complexes as building blocks for constructing new photofunctional molecules.

1.2 Overview of this Thesis

In Chap. 2, syntheses of various hetero-multinuclear complexes using the Mizoroki–Heck reaction are reported. Photofunctional mononuclear complexes possessing either vinyl groups or bromo groups were successfully connected in a single step. The photophysical properties and photocatalytic abilities of the synthesized trinuclear complexes are also investigated in this chapter.

Chapter 3 describes a further investigation on more suitable reaction conditions for the Mizoroki–Heck reactions, especially the effects of air and used phosphine ligands on the coupling reactions. I proposed two types of suitable reaction conditions for the Mizoroki–Heck reaction. Using a combination of these two conditions, hetero-trinuclear complexes, which have three different complex units, could be successfully obtained. The Mizoroki–Heck reactions were also combined with a homo-coupling reaction, which selectively proceeded between metal complexes with bromo groups under similar reaction conditions to synthesize various novel photofunctional multinuclear complexes.

Photochemical hydrogenation reactions of the synthesized multinuclear complexes in Chaps. 2 and 3 are reported in Chap. 4. This hydrogenation reaction could be applied for various complexes with not only vinylene linkers but also ethynylene linkers which were produced by using the Sonogashira coupling reactions. Some of the products of this hydrogenation reaction showed higher photocatalytic abilities compared to the corresponding complexes with unsaturated bonds.

Chapter 5 describes both synthetic methods for introducing the bromo groups into ring-shaped multinuclear Re(I) complexes and for connecting the Re(I) rings possessing bromo groups with another Re(I) ring or with a different kind of mononuclear complexes having a vinyl group. The synthesized complexes showed high light-harvesting ability owing to antenna effects of the Re(I) ring units and rapid intramolecular energy transfer.

References

1. Amadelli R, Argazzi R, Bignozzi CA, Scandola F (1990) Design of antenna-sensitizer polynuclear complexes. Sensitization of titanium dioxide with $[Ru(bpy)_2(CN)_2]_2Ru(bpy(COO)_2)_2{}^{2-}$. J Am Chem Soc 112:7099
2. Hwang I-W, Ko DM, Ahn TK, Yoon ZS, Kim D, Peng X, Aratani N, Osuka A (2005) Excitation energy migration in a dodecameric porphyrin wheel. J Phys Chem B 109:8643
3. Kaveevivitchai N, Chitta R, Zong R, El Ojaimi M, Thummel RP (2012) A molecular light-driven water oxidation catalyst. J Am Chem Soc 134:10721
4. Li F, Jiang Y, Zhang B, Huang F, Gao Y, Sun L (2012) Towards a solar fuel device: light-driven water oxidation catalyzed by a supramolecular assembly. Angew Chem Int Ed 51:2417
5. Ozawa H, Sakai K (2011) Photo-hydrogen-evolving molecular devices driving visible-light-induced water reduction into molecular hydrogen: structure-activity relationship and reaction mechanism. Chem Commun 47:2227
6. Arachchige SM, Brown JR, Chang E, Jain A, Zigler DF, Rangan K, Brewer KJ (2009) Design considerations for a system for photocatalytic hydrogen production from water employing mixed-metal photochemical molecular devices for photoinitiated electron collection. Inorg Chem 48:1899
7. Stoll T, Gennari M, Fortage J, Castillo CE, Rebarz M, Sliwa M, Poizat O, Odobel F, Deronzier A, Collomb M-N (2014) An efficient Ru^{II}–Rh^{III}–Ru^{II} polypyridyl photocatalyst for visible-light-driven hydrogen production in aqueous solution. Angew Chem Int Ed 53:1654
8. Koike K, Naito S, Sato S, Tamaki Y, Ishitani O (2009) Architecture of supramolecular metal complexes for photocatalytic CO_2 reduction III: effects of length of alkyl chain connecting photosensitizer to catalyst. J Photochem Photobiol A 207:109
9. Schneider J, Vuong KQ, Calladine JA, Sun X-Z, Whitwood AC, George MW, Perutz RN (2011) Photochemistry and photophysics of a Pd(II) metalloporphyrin: Re(I) tricarbonyl bipyridine molecular dyad and its activity toward the photoreduction of CO_2 to CO. Inorg Chem 50:118777
10. Tamaki Y, Watanabe K, Koike K, Inoue H, Morimoto T, Ishitani O (2012) Development of highly efficient supramolecular CO_2 reduction photocatalysts with high turnover frequency and durability. Faraday Discuss 155:115
11. Tamaki Y, Koike K, Morimoto T, Ishitani O (2013) Substantial improvement in the efficiency and durability of a photocatalyst for carbon dioxide reduction using a benzoimidazole derivative as an electron donor. J Catal 304:22
12. Gholamkhass B, Mametsuka H, Koike K, Tanabe T, Furue M, Ishitani O (2005) Architecture of supramolecular metal complexes for photocatalytic CO_2 reduction: Ruthenium-rhenium bi- and tetranuclear complexes. Inorg Chem 44:2326
13. Kato E, Takeda H, Koike K, Ohkubo K, Ishitani O (2015) Ru(II)–Re(I) binuclear photocatalysts connected by $-CH_2XCH_2-$ ($X = O, S, CH_2$) for CO_2 reduction. Chem Sci 6:3003

14. Balzani V, Campagna S, Denti G, Juris A, Serroni S, Venturi M (1998) Designing dendrimers based on transition-metal complexes. Light-harvesting properties and predetermined redox patterns. Acc Chem Res 31:26
15. Tamaki Y, Morimoto T, Koike K, Ishitani O (2012) Photocatalytic CO_2 reduction with high turnover frequency and selectivity of formic acid formation using Ru(II) multinuclear complexes. Proc Natl Acad Sci U S A 109:15673
16. Tamaki Y, Koike K, Morimoto T, Yamazaki Y, Ishitani O (2013) Red-light-driven photocatalytic reduction of CO_2 using Os(II)–Re(I) supramolecular complexes. Inorg Chem 52:11902
17. Sato S, Koike K, Inoue H, Ishitani O (2007) Highly efficient supramolecular photocatalysts for CO_2 reduction using visible light. Photochem Photobiol Sci 6:454
18. Kou Y, Nakatani S, Sunagawa G, Tachikawa Y, Masui D, Shimada T, Takagi S, Tryk DA, Nabetani Y, Tachibana H, Inoue H (2014) Visible light-induced reduction of carbon dioxide sensitized by a porphyrin–rhenium dyad metal complex on p-type semiconducting NiO as the reduction terminal end of an artificial photosynthetic system. J Catal 310:57
19. Bian Z-Y, Wang H, Fu W-F, Li L, Ding A-Z (2012) Two bifunctional Ru^{II}/Re^{I} photocatalysts for CO_2 reduction: a spectroscopic, photocatalytic, and computational study. Polyhedron 32:78
20. Kitagawa Y, Ogasawara S, Kosumi D, Hashimoto H, Tamiaki H (2015) Photophysical properties of chlorophyll derivatives linked with rhenium bipyridine complexes. Bull Chem Soc Jpn 88:346
21. Kitagawa Y, Takeda H, Ohashi K, Asatani T, Kosumi D, Hashimoto H, Ishitani O, Tamiaki H (2014) Photochemical reduction of CO_2 with red light using synthetic chlorophyll-rhenium bipyridine dyad. Chem Lett 43:1383
22. Nakada A, Koike K, Nakashima T, Morimoto T, Ishitani O (2015) Photocatalytic CO_2 reduction to formic acid using a Ru(II)-Re(I) supramolecular complex in an aqueous solution. Inorg Chem 54:1800
23. Morimoto T, Nishiura C, Tanaka M, Rohacova J, Nakagawa Y, Funada Y, Koike K, Yamamoto Y, Shishido S, Kojima T, Saeki T, Ozeki T, Ishitani O (2013) Ring-shaped Re(I) multinuclear complexes with unique photofunctional properties. J Am Chem Soc 135:13266
24. Rohacova J, Sekine A, Kawano T, Tamari S, Ishitani O (2015) Trinuclear and tetranuclear Re(I) rings connected with phenylene, vinylene, and ethynylene chains: synthesis, photophysics, and redox properties. Inorg Chem 54:8769
25. Asatani T, Nakagawa Y, Funada Y, Sawa S, Takeda H, Morimoto T, Koike K, Ishitani O (2014) Ring-shaped rhenium(I) multinuclear complexes: improved synthesis and photoinduced multielectron accumulation. Inorg Chem 53:7170
26. Connors PJ, Tzalis D, Dunnick AL, Tor Y (1998) Coordination compounds as building blocks: single-step synthesis of heteronuclear multimetallic complexes containing Ru^{II} and Os^{II}. Inorg Chem 37:1121
27. Tzalis D, Tor Y (1996) Coordination compounds as building blocks: single-step synthesis of multi-ruthenium(II) complexes. Chem. Commun 1043
28. Goeb S, De Nicola A, Ziessel R (2005) Controlled synthesis of multinuclear metal complex arrays by cross coupling of coordinated ligands. J Org Chem 70:6802
29. Arm KJ, Williams JAG (2006) A cross-coupling strategy for the synthesis of dimetallic assemblies containing mixed bipyridine-terpyridine bridging ligands: luminescence and energy transfer properties. Dalton Trans 2172
30. Whittle VL, Williams JAG (2009) Cyclometallated, bis-terdentate iridium complexes as linearly expandable cores for the construction of multimetallic assemblies. Dalton Trans 3929
31. Welter S, Salluce N, Belser P, Groeneveld M, De Cola L (2005) Photoinduced electronic energy transfer in modular, conjugated, dinuclear Ru(II)/Os(II) complexes. Coord Chem Rev 249:1360
32. Cassidy L, Horn S, Cleary L, Halpin Y, Browne WR, Vos JG (2009) Synthesis of asymmetric supramolecular compounds using a Ni(0) catalysed homo-coupling approach. Dalton Trans 3923
33. Bian Z-Y, Chi S-M, Li L, Fu W (2010) Conjugation effect of the bridging ligand on the CO_2 reduction properties in difunctional photocatalysts. Dalton Trans 39:7884

Chapter 2
Synthesis of Multinuclear Complexes Using the Mizoroki–Heck Reaction

Abstract An integration of photofunctional metal complexes is one of the useful strategies to develop multifarious photofunctional metal complexes. In particular, hetero-multinuclear complexes often show an enhancement in photoluminescent properties, photosensitizing abilities, and photocatalytic abilities due to the efficient intramolecular electron and/or excitation energy transfer. In recent years, several C–C coupling reactions, which have been frequently used in organic chemistry, were applied for connecting transition metal complexes. In this chapter, I focused on the Mizoroki–Heck reaction, which has few reports on application to integration of metal complexes. Various photofunctional metal complexes with functional groups, i.e., bromo and vinyl groups, were integrated into hetero-multinuclear complexes using the Mizoroki–Heck reaction under suitable reaction conditions. The obtained trinuclear complexes absorb a wide range of visible light and have long emission lifetimes and the photocatalytic ability for CO_2 reduction.

Keywords Mizoroki–Heck reaction · Photofunctional multinuclear complexes · Building block approach · Intramolecular energy transfer

2.1 Introduction

The construction of multinuclear complexes with various photofunctional metal complexes is one of the useful strategies for developing efficient photocatalysts [1–11] and light-harvesting systems [12] owing to rapid intramolecular electron transfer and/or energy transfer. Most of the reported photofunctional multinuclear metal complexes were synthesized using stepwise coordination of metal complexes to multi-dentate ligands, so-called "bridging ligands" [13]. In other words, this synthetic method requires a 1:1 combination of a metal complex having free coordinating sites in one side of the bridging ligand as a "*metal complex as ligand*", and another metal complex possessing easily replaceable ligands to accept coordination of the *metal complex as ligand*, as a "*metal complex as metal*". Although repeating these processes enables us to synthesize various multinuclear complexes, this synthetic method causes some problems, such as limitations on targeted molecules and product selectivity, as described in Chap. 1.

Y. Yamazaki, *Development of Synthetic Methods for Novel Photofunctional Multinuclear Complexes*, Springer Theses, https://doi.org/10.1007/978-981-16-7148-7_2

It has been recently reported that various coupling reactions, i.e., Sonogashira coupling [14–16], olefin metathesis [17, 18], homo-coupling reactions [19], and Suzuki–Miyaura coupling [20–22], are applicable for the construction of multinuclear complexes, and some of the problems derived from the ordinary synthetic method can be overcome by using these reactions. In these synthetic methods, functional groups, e.g., bromo groups, olefin groups, ethynyl groups, and boronic–acid substituents, are introduced into the ligands on metal complexes as reactive sites, and the complexes are coupled each other to give a single multinuclear complex with relatively high selectivity and yields.

Herein I report the first use of the Mizoroki–Heck reaction [23, 24] in a coupling reaction with different types of emissive metal complexes (Ru(II), Ir(III), and Re(I) complexes with one or more diimine ligands) to give multinuclear complexes with photofunctional properties. These multinuclear complexes show strong absorption ability and long emission lifetimes, which are useful properties for utilizing as light absorbers for such as light-harvesting systems and dye-sensitized solar cells. A multinuclear complex having one Re(I) catalyst unit with two photosensitizer units, which cannot be selectively synthesized with ordinary synthetic methods, was successfully synthesized, and it showed photocatalytic ability for CO_2 reduction. However, the reaction selectivity of reduction products was not high, probably due to decomposition reactions of the photosensitizer units derived from the low reduction power of the catalyst unit. The photocatalytic ability of this complex could be improved by using the hydrogenation reaction as described in Chap. 4.

2.2 Results and Discussion

2.2.1 Synthesis of a Re(I) Dinuclear Complex Using the Mizoroki–Heck Reaction

I chose two types of Re(I) complexes as building blocks (Eq. 2.1) to check if the Mizoroki–Heck reaction is able to connect different photofunctional metal complexes. An acetonitrile (MeCN) solution (4 mL) containing the two Re(I) complexes, one with a bromo group (**[Re(Br,Et)]**$^+$, 23 μmol) and the other with a vinyl group (**[Re(C = C,OEt)]**$^+$, 23 μmol), was heated in the presence of palladium acetate ($Pd(OAc)_2$, 4.5 μmol), triphenylphosphine (PPh_3, 9.2 μmol) and sodium acetate (AcONa, 140 μmol) at 75 °C under an Ar atmosphere for one day. Ar gas was pumped into the reaction mixture during synthesis. The solvent was evaporated once the reaction reached completion. A MeCN solution (4 mL) of $Pd(OAc)_2$ (4.5 μmol) and PPh_3 (9.2 μmol) was then added to the residue along with a small amount of air; this addition of air was needed to achieve a good yield [25]. The solution was heated for one day. Reaction progression was monitored by size exclusion chromatography (SEC) [26]. The SEC charts are illustrated in Fig. 2.1. The new peak possessing a less retention time, compared to the starting complexes, was observed. This result showed

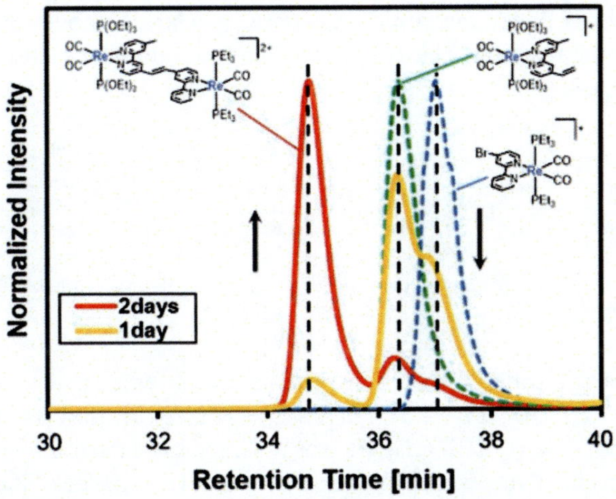

Fig. 2.1 SEC charts measured during the synthesis of **[Re(OEt)Re(Et)]$^{2+}$**. The detection wavelength was 360 nm

that some new species, of which molecular size is larger than starting materials, were produced during the reaction, especially after charging reagents with some air. Electrospray ionization mass spectrometry (ESI–MS) analysis of the crude products indicated three major peaks attributed to **[Re(OEt)Re(Et)]$^{2+}$** as the main product, **[Re(C = C,OEt)]$^{+}$** and [Re(bpy)(CO)$_2$(PEt$_3$)$_2$]$^{+}$ (bpy = 2,2′-bipyridine), which should be produced via debromination of **[Re(Br,Et)]$^{+}$** (Fig. 2.2). The target dinuclear Re(I) complex (**[Re(OEt)Re(Et)]$^{2+}$**) was chromatographically isolated using silica gel and SP Sephadex as adsorbents, giving an isolated yield of 70%. Note that the reaction

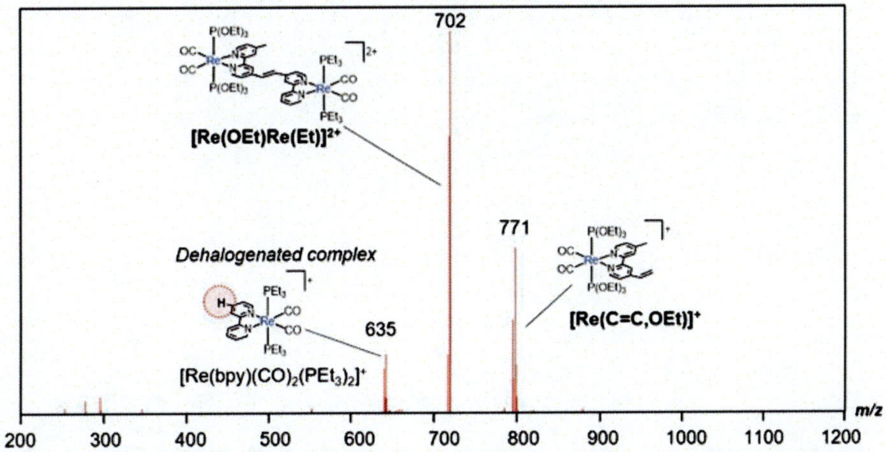

Fig. 2.2 ESI–MS spectrum of crude products measured after the synthesis of **[Re(OEt)Re(Et)]$^{2+}$**. The mobile phase was MeCN

was much slower under anaerobic conditions, as a small amount of air was required to accelerate the reaction. The effect of air on the coupling reactions is discussed in detail in Chap. 3.

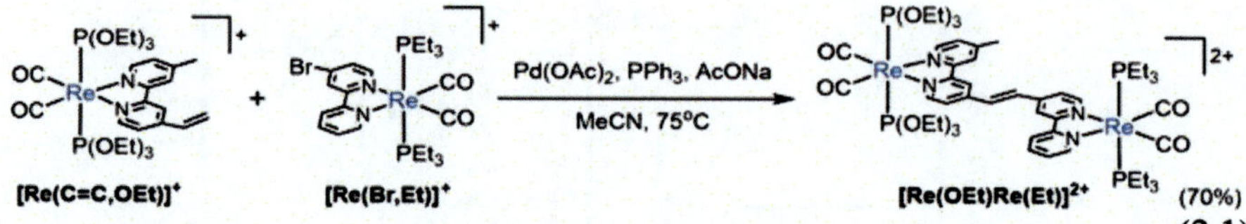

(2.1)

A single peak attributed to olefinic protons was observed in the ^{1}H NMR spectrum of **[Re(OEt)Re(Et)]$^{2+}$** (Fig. 2.3). This suggests that either the cis or trans isomer was formed. Product **[Re(OEt)Re(Et)]$^{2+}$** was assumed to be the trans isomer because the Re(I) units on either side were bulky. Such low-field resonance of the olefinic protons ($\delta = 8.07$) was similar to the resonance seen for *trans*-1,2-bis(4-(4′-methyl)-2,2′-bipyridine)ethene [27]. The high-resolution mass spectrum and ^{31}P NMR spectrum clearly showed that neither the $P(OEt)_3$ nor the PEt_3 ligands on the Re(I) units have been substituted with the coupling reagents.

It should be emphasized that the selective synthesis of only one type of dinuclear complex with an asymmetric bridging ligand, such as **[Re(OEt)Re(Et)]$^{2+}$**, is difficult using ordinary methods, which possibly results in a mixture of four types of dinuclear complexes, i.e., two types of complexes with identical phosphine ligands in both units

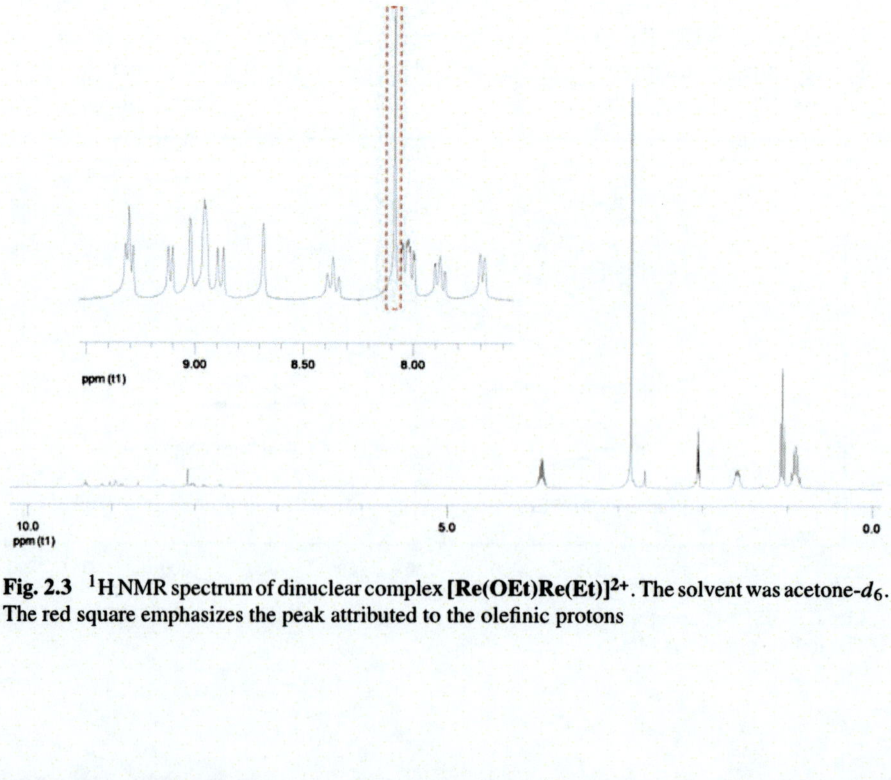

Fig. 2.3 ^{1}H NMR spectrum of dinuclear complex **[Re(OEt)Re(Et)]$^{2+}$**. The solvent was acetone-d_6. The red square emphasizes the peak attributed to the olefinic protons

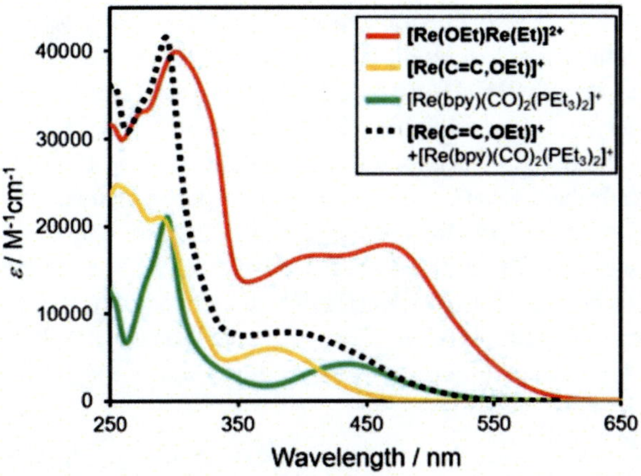

Fig. 2.4 UV–Vis absorption spectra of **$[Re(OEt)Re(Et)]^{2+}$** (red), **$[Re(C = C,OEt)]^{+}$** (yellow), **$[Re(bpy)(CO)_2(PEt_3)_2]^{+}$** (green) measured in MeCN and 1:1 summation spectrum of **$[Re(C = C,OEt)]^{+}$** and **$[Re(bpy)(CO)_2(PEt_3)_2]^{+}$** (dotted line)

and two structural isomers of complexes with two different phosphine ligands in each unit.

The absorption spectrum of **$[Re(OEt)Re(Et)]^{2+}$** and spectra of the corresponding mononuclear complexes are shown in Fig. 2.4. The singlet metal-to-ligand charge transfer (^{1}MLCT) absorption bands were red-shifted, and the molar extinction coefficients of the bands were increased, compared to the summation spectrum of corresponding mononuclear complexes because of the large π conjugation systems; this complex can utilize a wide range of visible light.

2.2.2 *Synthesis of Various Trinuclear Complexes Using the Mizoroki–Heck Reaction*

A similar reaction could successfully be used to synthesize various trinuclear complexes. In the center of each of these products was a Re(I) complex with a bpy ligand and two other metal complexes as substituents (Eq. 2.2). A Re(I) biscarbonyl complex with two bromo groups, $[Re^{I}(4,4'\text{-dibromo-bpy})(CO)_2(PPh_3)_2]^{+}$ (**$[Re(Br_2,Ph)]^{+}$**), reacted with various photofunctional complexes possessing a vinyl group, i.e., **$[Re(C = C,OEt)]^{+}$**, $[Ru^{II}(bpy)_2(vbpy)]^{2+}$ (**$[Ru(C = C,bpy)]^{2+}$**, vbpy = 4-methyl-4′-vinyl-bpy), $[Ru^{II}(dmb)_2(vbpy)]^{2+}$ (**$[Ru(C = C,dmb)]^{2+}$**, dmb = 4,4′-dimethyl-bpy) and $[Ir^{III}(ppy)(vbpy)]^{+}$ (**$[Ir(C = C,ppy)]^{+}$**, ppy = 2-phenylpyridine). This was performed under the same reaction conditions as were used to synthesize **$[Re(OEt)Re(Et)]^{2+}$**, with the exception of repeating both the heating and addition of the MeCN solution of $Pd(OAc)_2$ and PPh_3 twice, in the case of **$[Ru(bpy)_2Re(Ph)]^{5+}$** and **$[Ir(ppy)_2Re(Ph)]^{3+}$**. The trinuclear complexes could be separated using size exclusion chromatography (SEC) or ion-exchange chromatography. The trinuclear complexes with Ru(II) or Re(I) in both edge units, i.e., **[Re(OEt)$_2$Re(Ph)]$^{3+}$**, **[Ru(bpy)$_2$Re(Ph)]$^{5+}$** and **[Ru(dmb)$_2$Re(Ph)]$^{5+}$**, were obtained in 31%, 41% and 32% yields, respectively. ESI–MS analysis of the solution after the reaction between [Re(4,4′-dibromo-bpy)(CO)$_2${P(*p*-FC$_6$H$_4$)$_3$}$_2$]$^+$ (**[Re(Br$_2$,FPh)]$^+$**) and **[Ru(C = C,dmb)]$^{2+}$** was performed and clearly showed that the phosphine ligands on the Re(I) unit were not substituted by the coupling reagents. The isolated yield of **[Ru(dmb)$_2$Re(FPh)]$^{5+}$** was 25%. The isolated yield was relatively lower (8%) for **[Ir(ppy)$_2$Re(Ph)]$^{3+}$**, which had Ir(III) in both edge units, as the reaction between **[Ir(C = C,ppy)]$^+$** and **[Re(Br$_2$,Ph)]$^+$** was much slower compared to the other reactions; therefore, conversion of the starting complexes was low (Fig. 2.5).

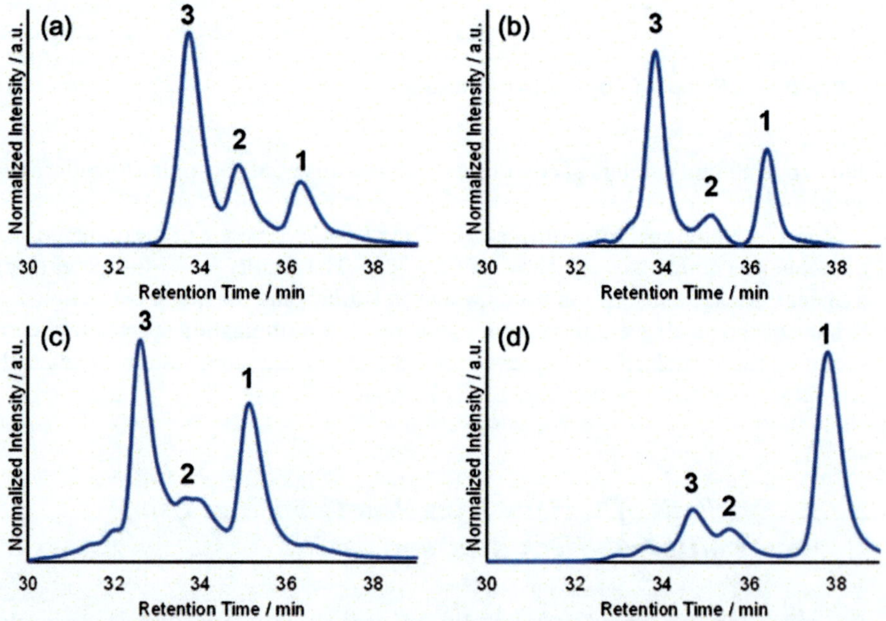

Fig. 2.5 SEC charts of the reaction mixture measured after the coupling reaction. **a** **[Re(OEt)$_2$Re(Ph)]$^{3+}$**, **b** **[Ru(bpy)$_2$Re(Ph)]$^{5+}$**, **c** **[Ru(dmb)$_2$Re(Ph)]$^{5+}$** and **d** **[Ir(ppy)$_2$Re(Ph)]$^{3+}$**. **1**, **2** and **3** in each figure represent the peaks attributed to mononuclear complexes, dinuclear complexes and trinuclear complexes, respectively

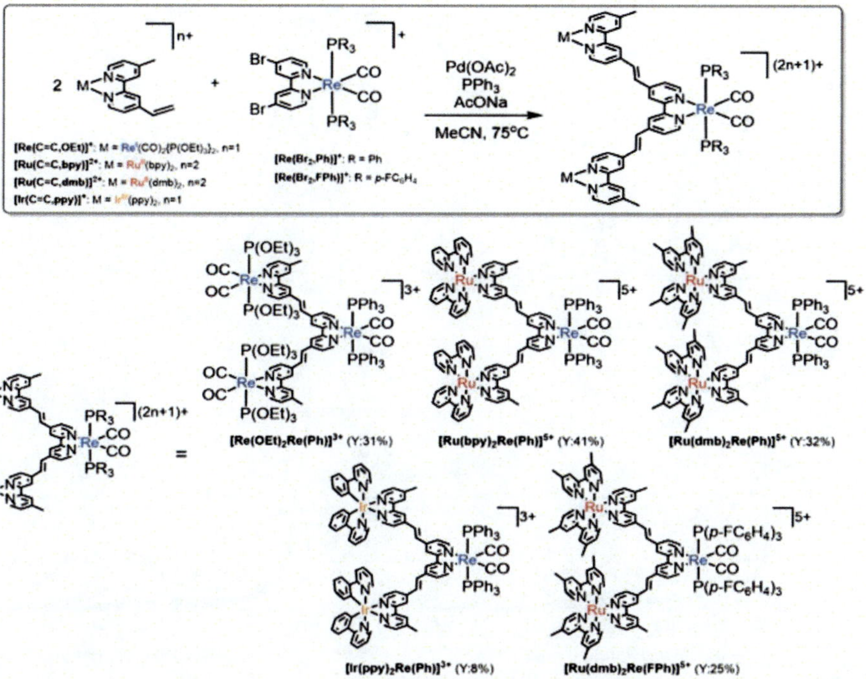

(2.2)

As is described above, the complexes that were synthesized in this research will not be obtained selectively using ordinary stepwise methods because mixtures of isomers with the metals in different units will be produced. Isolating the target complex from such a mixture is often difficult because of the similarities between the molecular sizes, charges, and solubility of the isomeric complexes.

2.2.3 Photophysical Properties and Photocatalytic Abilities of the Trinuclear Complexes

2.2.3.1 Absorption Properties

UV–Vis absorption spectra of the synthesized trinuclear complexes **[Ru(dmb)$_2$Re(Ph)]$^{5+}$**, **[Re(OEt)$_2$Re(Ph)]$^{3+}$**, **[Ir(ppy)$_2$Re(Ph)]$^{3+}$**, **[Ru(dmb)$_2$Re(Ph)]$^{5+}$** and **[Ru(dmb)$_2$Re(FPh)]$^{5+}$** are shown with those of the corresponding mononuclear complexes in Fig. 2.6. The ^{1}MLCT absorption band of **[Ru(dmb)$_2$Re(Ph)]$^{5+}$** was observed between 370 and 570 nm and was red-shifted compared to the ^{1}MLCT absorption band of **[Ru(C = C,dmb)]$^{2+}$** (Fig. 2.6a). This was most likely caused by the extension of the π conjugation in the bridging ligand, decreasing the energy level of the $\pi*$ orbital [28]. In addition, the molar extinction

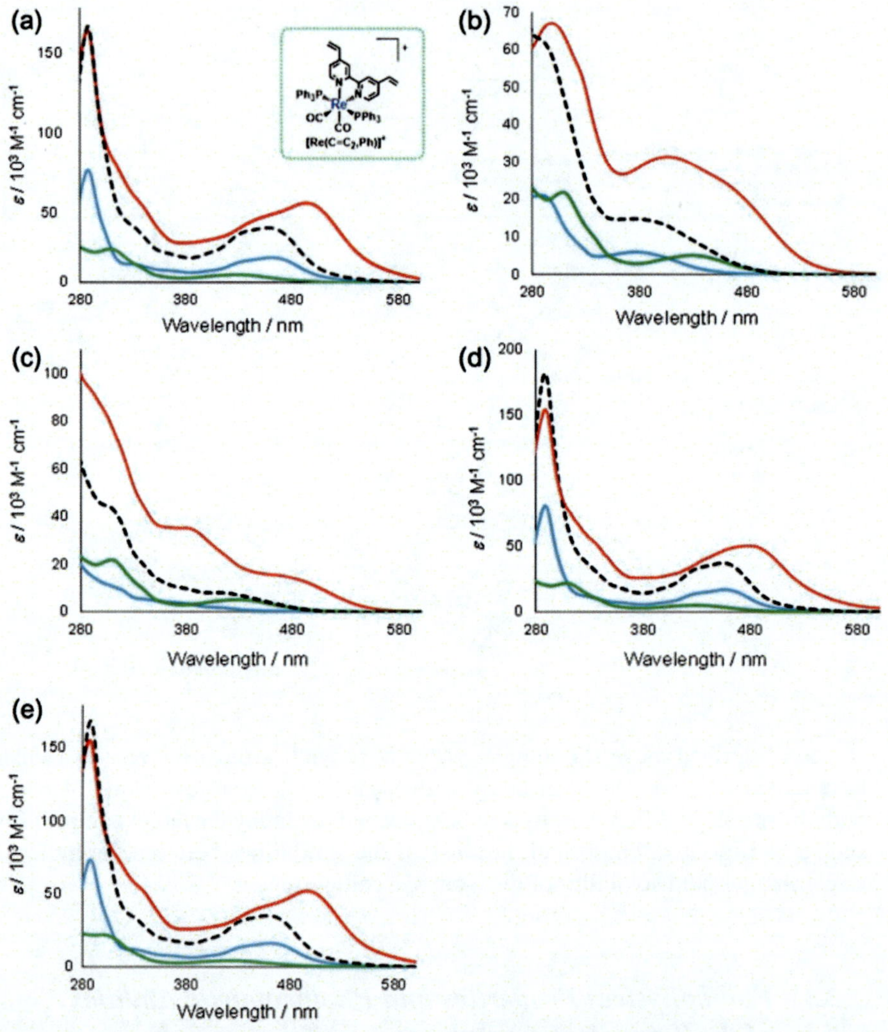

Fig. 2.6 UV–Vis absorption spectra of obtained trinuclear complexes and corresponding mononuclear complexes. The solvent was MeCN. **a** **$[Ru(dmb)_2Re(Ph)]^{5+}$** (red), **$[Ru(C = C,dmb)]^{2+}$** (blue), **$[Re(C = C_2,Ph)]^{+}$** (green) and the 2:1 summation spectrum of **$[Ru(C = C,dmb)]^{2+}$** and **$[Re(C = C_2,Ph)]^{+}$** (dotted line). **b** **$[Re(OEt)_2Re(Ph)]^{3+}$** (red), **$[Re(C = C,OEt)]^{+}$** (blue), **$[Re(C = C_2,Ph)]^{+}$** (green) and the 2:1 summation spectrum of **$[Re(C = C,OEt)]^{+}$** and **$[Re(C = C_2,Ph)]^{+}$** (dotted line). **c** **$[Ir(ppy)_2Re(Ph)]^{3+}$** (red), **$[Ir(C = C,ppy)]^{+}$** (blue), **$[Re(C = C_2,Ph)]^{+}$** (green) and the 2:1 summation spectrum of **$[Ir(C = C,ppy)]^{+}$** and **$[Re(C = C_2,Ph)]^{+}$** (dotted line). **d** **$[Ru(bpy)_2Re(Ph)]^{5+}$** (red), **$[Ru(C = C,bpy)]^{2+}$** (blue), **$[Re(C = C_2,Ph)]^{+}$** (green) and the 2:1 summation spectrum of **$[Ru(C = C,bpy)]^{2+}$** and **$[Re(C = C_2,Ph)]^{+}$** (dotted line). **e** **$[Ru(dmb)_2Re(FPh)]^{5+}$** (red), **$[Ru(C = C,dmb)]^{2+}$** (blue), **$[Re(C = C_2,FPh)]^{+}$** (green) and the 2:1 summation spectrum of **$[Ru(C = C,dmb)]^{2+}$** and **$[Re(C = C_2,FPh)]^{+}$** (dotted line)

coefficient of the ^{1}MLCT absorption band was higher for **[Ru(dmb)$_2$Re(Ph)]$^{5+}$** than for the corresponding mononuclear complexes. This suggests that the transition has more π–π* character. The π–π* absorption band of **[Ru(dmb)$_2$Re(Ph)]$^{5+}$** with $\lambda_{max} = 288$ nm had a shoulder at longer wavelengths, which was not observed in the summation spectra of the corresponding mononuclear complexes (dotted line in Fig. 2.6a). This shoulder can be attributed to the π–π* transition in the bridging ligand with a wide π conjugation system. In the case of the other trinuclear complexes, similar phenomena were observed, i.e., red-shifts of both the π–π* and ^{1}MLCT absorption bands and increase of the molar extinction coefficients of the ^{1}MLCT absorption bands (Table 2.1).

2.2.3.2 Emission Properties

All trinuclear complexes displayed emission (Figs. 2.7 and 2.8). The emission for each trinuclear complex was red-shifted relative to the corresponding mononuclear complexes and the emission quantum yield (Φ_{em}) for each trinuclear complex was much lower than those of the corresponding mononuclear complexes (Table 2.1).

The emission decay for each trinuclear complex could be fitted using a double exponential function (Fig. 2.9). The emission lifetimes (τ_{em}) were similar to the corresponding mononuclear complexes, though the Φ_{em} were much lower; therefore, the radiative decay of the excited states of the trinuclear complexes should be slower than those of the mononuclear complexes with respect to the following equations, which show the relationships among Φ_{em}, τ_{em} and the radiative and nonradiative decay constants (k_r and k_{nr}, Eqs. 2.3 and 2.4).

$$\Phi_{em} = k_r/(k_r + k_{nr}) \tag{2.3}$$

$$\tau_{em} = 1/(k_r + k_{nr}) \tag{2.4}$$

This also indicates that the lowest ^{3}MLCT excited states of the trinuclear complexes contained higher π–π* characteristics than those of the mononuclear complexes. It has been reported that an increase in π–π* character of the ^{3}MLCT excited states of transient metal complexes induces slower radiative decay [27, 29].

2.2.3.3 Temperature Dependence of Emission Lifetime and Time-Resolved Emission Spectra

For a more detailed investigation of the excited state of the trinuclear complexes, time-resolved emission spectra and temperature dependence of the emission decay of **[Ru(dmb)$_2$Re(Ph)]$^{5+}$** were measured as a typical example. Figure 2.10a shows the emission decays measured at various temperatures (243–333 K) after excitation at $\lambda_{ex} = 456$ nm, where much weaker temperature dependence was observed compared to $[Ru(dmb)_3]^{2+}$ (Figs. 2.10b, 2.11). It has been reported that emission decays of many

Table 2.1 Photophysical properties of the trinuclear complexes and corresponding mononuclear complexes[a]

Entry	Complex	λ_{abs}/nm ($\varepsilon/10^3$ M^{-1} cm^{-1})	$\lambda_{em}{}^{max}$/nm	Φ_{em}	τ_1/ns^d (A/%)	τ_2/ns^d (A/%)
1	**$[Re(OEt)_2Re(Ph)]^{3+}$**	406 (31.3), 476 (sh, 20.8), 299 (65.4)	714[c]	0.01[c]	140 (56)[c]	771 (44)[c]
2	**$[Ru(bpy)_2Re(Ph)]^{5+}$**	481 (50.6), 341 (sh, 47.3), 287 (162)	710[b]	0.02[b]	674 (34)[b]	1424 (66)[b]
3	**$[Ru(dmb)_2Re(Ph)]^{5+}$**	494 (52.1), 341 (sh, 48.9), 288(150)	738[b]	0.02[b]	610 (18)[b]	1310 (82)[b]
4	**$[Ir(ppy)_2Re(Ph)]^{3+}$**	474 (14.9), 383 (35.0), 257 (120)	685[b]	0.01[c]	104 (82)[c]	784 (18)[c]
5	**$[Ru(dmb)_2Re(FPh)]^{5+}$**	495 (52.3), 341 (sh, 50.7), 287 (156)	743[b]	0.03[b]	471 (6)[b]	1112 (94)[b]
6	**$[Re(C = C,OEt)]^{+}$**	376 (5.97), 290 (20.4)	635[c]	0.07[c]	326 (100)[c]	-
7	**$[Ru(C = C,bpy)]^{2+}$**	456 (17.0), 289 (82.1)	636[c]	0.14[b]	1551 (100)[b]	-
8	**$[Ru(C = C,dmb)]^{2+}$**	462 (16.3), 288 (73.1)	645[c]	0.10[b]	1376 (100)[b]	-
9	**$[Ir(C = C,ppy)]^{+}$**	384 (3.26), 304 (11.5)	617[c]	0.14[c]	200 (100)[c]	-
10	**$[Re(C = C_2,Ph)]^{+}$**	428 (5.38), 308 (21.9)	647[c]	0.08[c]	1561 (100)[c]	-
11	**$[Re(C = C_2,FPh)]^{+}$**	424 (4.29), 300 (22.2)	642[c]	0.07[c]	1409 (100)[c]	-

[a] Measured in MeCN at 25 °C. [b] Excitation wavelength: 456 nm. [c] Excitation wavelength: 444 nm. [d] Detected at each wavelength of the emission maximum

Ru(II) [30] trisdiimine and Re(I) [31] diimine phosphine complexes are strongly dependent on temperature, becoming faster at higher temperatures due to thermal interconversion from the lowest ^{3}MLCT excited state to the triplet ligand field excited state (^{3}LF), which causes rapid nonradiative decay (Fig. 2.12). Therefore, the formation process of the ^{3}LF should be suppressed in the case of **$[Ru(dmb)_2Re(Ph)]^{5+}$**, as the extension of π^* conjugation in the bridging ligand lowers the energy level of the ^{3}MLCT excited state. Suppression of this nonradiative decay process might be a

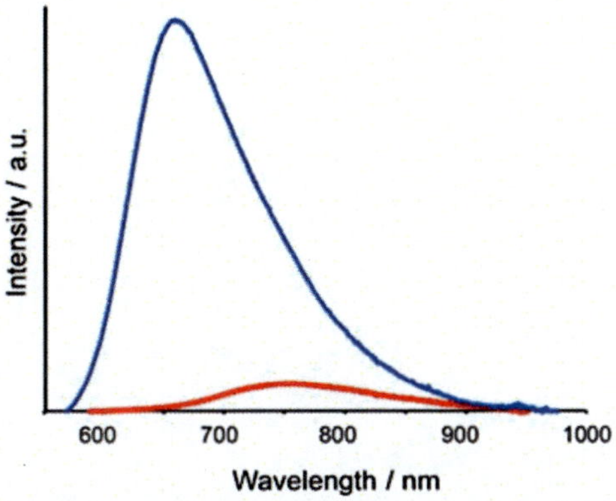

Fig. 2.7 Emission spectra of **[Ru(C = C,dmb)]$^{2+}$** (blue) and **[Ru(dmb)$_2$Re(Ph)]$^{5+}$** (red) measured in MeCN (excitation wavelength: 456 nm)

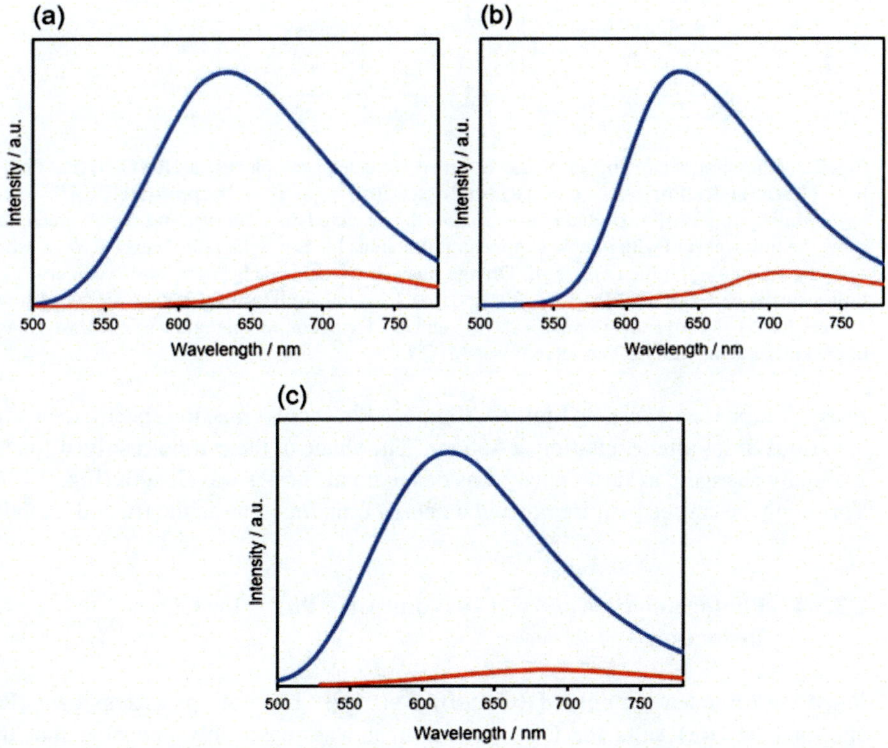

Fig. 2.8 Emission spectra of obtained trinuclear complexes and corresponding mononuclear complexes measured in MeCN. **a** **[Re(C = C,OEt)]$^{+}$** (blue) and **[Re(OEt)$_2$Re(Ph)]$^{3+}$** (red). The excitation wavelength was 444 nm. **b** **[Ru(C = C,bpy)]$^{2+}$** (blue) and **[Ru(bpy)$_2$Re(Ph)]$^{5+}$** (red). The excitation wavelength was 456 nm. **c** **[Ir(C = C,ppy)]$^{+}$** (blue) and **[Ir(ppy)$_2$Re(Ph)]$^{3+}$** (red). The excitation wavelength was 444 nm

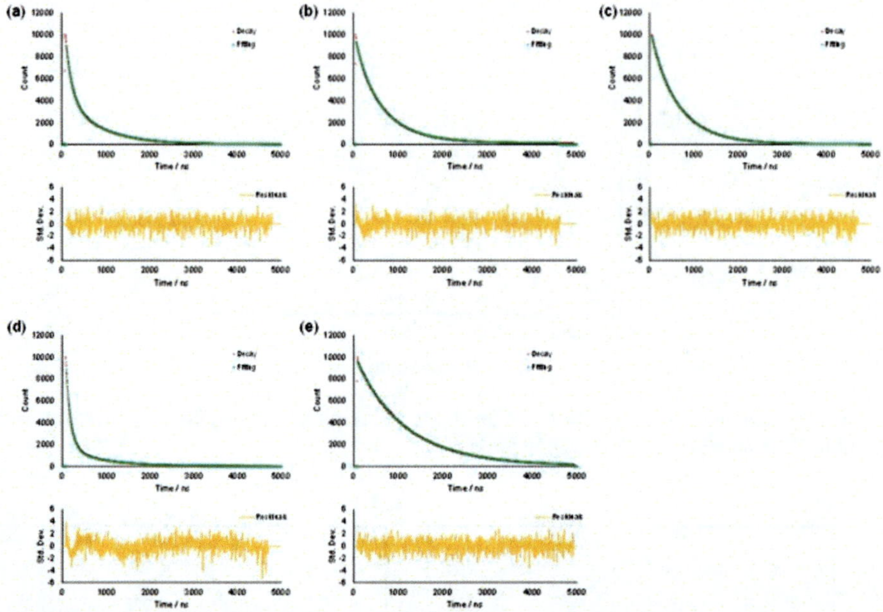

Fig. 2.9 Emission decay curves of the obtained trinuclear complexes. **a** **[Re(OEt)$_2$Re(Ph)]$^{3+}$**, **b** **[Ru(bpy)$_2$Re(Ph)]$^{5+}$**, **c** **[Ru(dmb)$_2$Re(Ph)]$^{5+}$**, **d** **[Ir(ppy)$_2$Re(Ph)]$^{3+}$** and **e** **[Ru(dmb)$_2$Re(FPh)]$^{5+}$**. Dotted red lines show the observed decay curves and dotted green lines show the fitting curves with double exponential functions [$A_1\exp(-t/\tau_1) + A_2\exp(-t/\tau_2)$, where τ_i is the emission lifetime ($\tau_1 < \tau_2$)]. Orange lines show residuals between observed curves and fitting curves (χ^2: (a) 0.98, (b) 1.02, (c) 1.00, (d) 1.24 and (e) 0.99). Excitation wavelength was 444 nm for (a) and (d) and 456 nm for (b), (c) and (e). Detection wavelength was each wavelength of the emission maximum. The solvent was MeCN

reason for the long emission lifetime. Figure 2.13a shows emission spectra measured at various times after excitation at 456 nm. The shape of these time-resolved spectra gradually changed, as faster decay was observed at shorter wavelength (Fig. 2.13b). This might be caused by intramolecular energy transfer between the Ru and Re units.

2.2.3.4 Photocatalytic Ability of [Ru(dmb)$_2$Re(Ph)]$^{5+}$ for CO_2 Reduction

Since the trinuclear complex **[Ru(dmb)$_2$Re(Ph)]$^{5+}$** has both photosensitizer (Ru) and catalyst (Re) units for CO_2 reduction, it was reasonably expected that this complex could act as a photocatalyst for CO_2 reduction. A mixture of DMF and triethanolamine (5:1 v/v) containing 0.05 mM **[Ru(dmb)$_2$Re(Ph)]$^{5+}$** and 0.1 M 1,3-dimethyl-2-phenyl-2,3-dihydro-1*H*-benzo[*d*]imidazole (BIH) was irradiated at $\lambda_{ex} = 600$ nm for 22 h. Selective formation of CO was observed, and the turnover number (TON_{CO}) reached 115. This TON_{CO} was higher than that found using a Ru(II)–Re(I) dinuclear complex with an N^N–CH_2–CH_2–N^N bridging ligand (N^N = 4′-methyl-bpy) ($TON_{CO} = 50$) under the same reaction conditions, which has been reported to

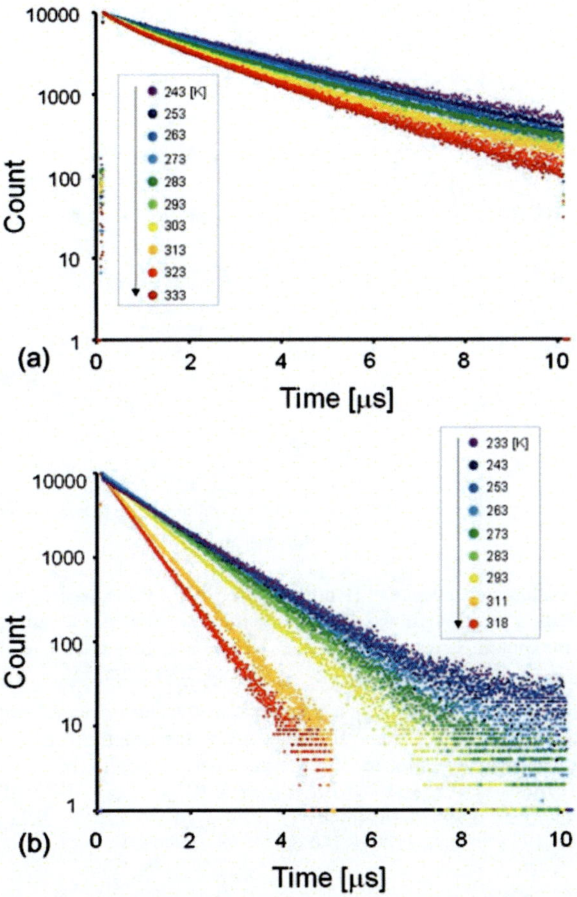

Fig. 2.10 **a** Temperature dependence of emission decay of **$[Ru(dmb)_2Re(Ph)]^{5+}$**. Decay curves were measured from 243 K (purple) to 333 K (dark red). The excitation wavelength was 456 nm and the detection wavelength was 738 nm. **b** Temperature dependence of emission decay of $[Ru(dmb)_3]^{2+}$. Decay curves were measured from 233 K (purple) to 318 K (red). The excitation wavelength was 456 nm and the detection wavelength was 639 nm. The solvent was MeCN

be one of the most efficient supramolecular photocatalysts for the reduction of CO_2 using higher-energy light such as 480 nm [8], because the absorption at $\lambda = 600$ nm is stronger. Under these reaction conditions, most Ru(II) moieties should be mainly excited (Fig. 2.6a). After a reductive quenching process of the excited state of the Ru(II) unit by BIH, intramolecular electron transfer should proceed from the reduced Ru(II) unit to the Re(I) unit, followed by reduction of CO_2 on the reduced Re(I) unit. However, the TON_{CO} is much lower than the reported systems [9]. The photocatalytic abilities of the $Ru(II)_2$-Re(I) trinuclear complexes and photodecomposition processes during photocatalytic reactions are discussed in detail in Chap. 4.

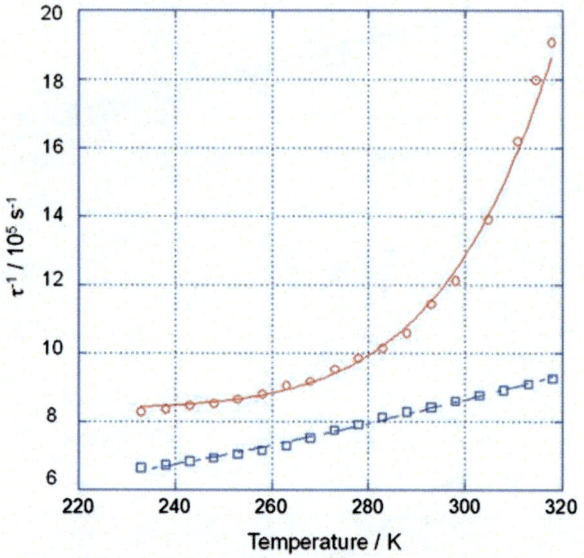

Fig. 2.11 Temperature dependence of **[Ru(dmb)$_2$Re(Ph)]$^{5+}$** (blue square) and [Ru(dmb)$_3$]$^{2+}$ (red circle). The excitation wavelength was 456 nm and the detection wavelength was each wavelength of the emission maximum. The solvent was MeCN. The blue line and the red line show the fitting curves based on the following equation: $\tau^{-1} = \frac{(k_r + k_{nr1}) + k_{nr2}\exp(-\Delta E/RT)}{1+\exp(-\Delta E/RT)}$ where k_r is a radiative decay constant from ^{3}MLCT state, k_{nr1} is a nonradiative decay constant from ^{3}MLCT state, k_{nr2} is a nonradiative decay constant from ^{3}LF state, ΔE is the difference between the energy level of ^{3}MLCT state and that of ^{3}LF state, R is the gas constant, and T is the absolute temperature. (**[Ru(dmb)$_2$Re(Ph)]$^{5+}$**; $k_r + k_{nr1}$: $4.7 \times 10^5 \pm 0.37 \times 10^5$ s^{-1}, k_{nr2}: $7.1 \times 10^6 \pm 1.2 \times 10^6$ s^{-1}, ΔE: 580 ± 59 cm^{-1}, R^2: 0.99554. [Ru(dmb)$_3$]$^{2+}$; $k_r + k_{nr1}$: $8.4 \times 10^5 \pm 0.12 \times 10^5$ s^{-1}, k_{nr2}: $1.3 \times 10^{12} \pm 0.92 \times 10^{12}$ s^{-1}, ΔE: 3100 ± 160 cm^{-1}, R^2: 0.99493)

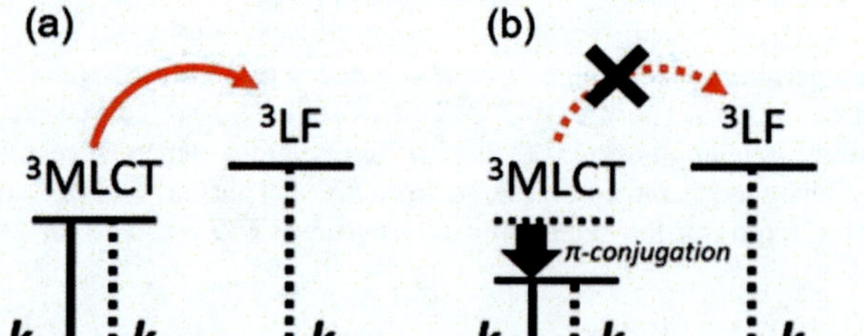

Fig. 2.12 Schematic Jablonski diagrams of **a** [Ru(dmb)$_3$]$^{2+}$ and **b [Ru(dmb)$_2$Re(Ph)]$^{5+}$**

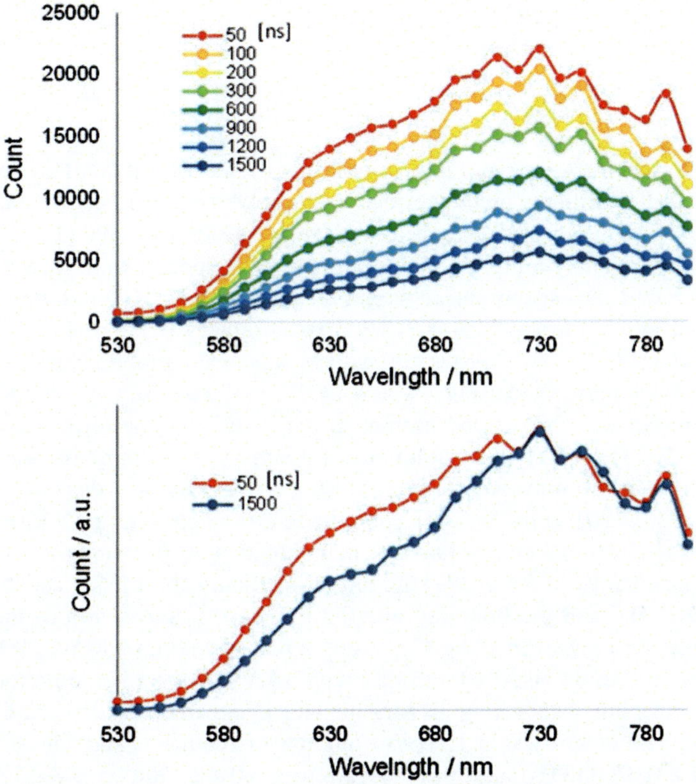

Fig. 2.13 **a** Transient emission spectra of **[Ru(dmb)$_2$Re(Ph)]$^{5+}$** (50–1500 ns after excitation) measured by single-photon counting methods and **b** normalised spectra at 730 nm (50 ns (red) and 1500 ns (blue) after excitation). The excitation wavelength was 444 nm. The solvent was MeCN

2.3 Conclusions

I successfully synthesized various trinuclear complexes from photofunctional complexes possessing vinyl or bromo groups as building blocks in a single step using the Mizoroki–Heck reaction. The obtained trinuclear complexes displayed strong absorption over a wide range of visible light and a long emission lifetime. **[Ru(dmb)$_2$Re(Ph)]$^{5+}$** functioned as a photocatalyst for CO_2 reduction. The obvious decomposition reaction during photocatalysis decrease the reaction selectivity and efficiency.

2.4 Experimental

2.4.1 General Procedures

^{1}H NMR spectra were measured using a JEOL AL400 (400 MHz) or an AL300 (300 MHz) instrument with the analyte dissolved in acetone-d_6, MeCN-d_3, or $CDCl_3$. ^{31}P NMR spectra were measured using a JEOL ECX400 (162 MHz) instrument with the analyte dissolved in acetone-d_6. IR spectra were measured using a JASCO FT/IR-610 spectrometer with a resolution of 1 cm^{-1} and with the analyte dissolved in dichloromethane. ESI mass spectrometry (MS) was performed using a Shimadzu LC–MS-2010A system, with MeCN as the mobile phase. ESI time-of-flight MS was performed using a Waters LCT Premier instrument, with MeCN as the mobile phase. Size Exclusion Chromatography (SEC) was performed using a pair of Shodex PROTEIN KW-402 J columns (300 mm long, 8.0 mm i.d.) with a KW-LG guard column (50 mm long, 6.0 mm i.d.), a JASCO 880–51 degasser, an 880-PU pump, an MD-2010 Plus UV–Vis photodiode-array detector, and a Rheodyne 7125 injector. The column temperature was maintained at 40 °C using a JASCO 860-CO oven. The eluent was a 1:1 (v/v) mixture of methanol and MeCN containing 0.5 M CH_3COONH_4, and the flow rate was 0.2 mL min^{-1}. Separation of the trinuclear complexes was achieved using SEC using a pair of Shodex PROTEIN KW-2002.5 columns (300 mm long, 20.0 mm i.d.) with a KW-LG guard column (50 mm long, 8.0 mm i.d.) and a recycling preparative HPLC apparatus with a JASCO 870-UV detector. The eluent was a 1:1 (v/v) mixture of methanol and MeCN containing 0.15 M CH_3COONH_4, and the flow rate was 6.0 mL min^{-1}. UV–Vis absorption spectra were recorded using a JASCO V-670 instrument.

2.4.2 Emission Measurements

Each compound was dissolved in MeCN and degassed using the freeze–pump–thaw method prior to measurement. Emission spectra were measured at 25 °C using either a JASCO FP-8600 spectrofluorometer or a Hamamatsu C9920-02 system. The absolute emission quantum yields were evaluated using the Hamamatsu C9920-02 system equipped with an integrating sphere and a multichannel photodetector (PMA-12). Emission lifetimes were obtained using a HORIBA TemPro fluorescence lifetime system with an emission monochromator. The excitation light source was a NanoLED-560 pulse lamp and the instrumental response time was less than 0.1 ns. All emission decays were fitted by single or double exponential functions within proper χ^2 (0.96–1.24). Transient emission spectra were measured by the point-to-point measurement method using a Horiba FluoroCube time-correlated single-photon counting system. The sensitivity of the system at each detection wavelength was calibrated using correction coefficients determined from the difference between

[Ru(4,4′-dimethyl-2–2′-bipyridine)$_3$]$^{2+}$ spectra measured with the JASCO FP-8600 spectrofluorometer and the HORIBA TemPro fluorescence lifetime system.

2.4.3 *Photocatalytic Reactions*

Photocatalytic reactions were performed in an 11 mL-cubic-quartz cell (with a light passage length of 1 cm) containing a 4 mL DMF–triethanolamine (TEOA) (5:1 v/v) solution of **[Ru(dmb)$_2$Re(Ph)]$^{5+}$** or the Ru(II)-Re(I) dinuclear complexes (0.05 mM) and BNAH (0.1 M) after purging with CO_2 for more than 20 min. The solution was irradiated with 600 nm monochromic light obtained from a Xe lamp with a band-pass filter (FWHM = 10 nm). The incident light intensity was 4.3×10^{-8} einstein/s. During the irradiation, the cell was cooled with a thermostatic bath. The gaseous reaction products (CO and H_2) were analyzed using a GC-TCD instrument (GL Science GC323). HCOOH was analyzed using a capillary electrophoresis system (Otsuka Electronics Co. CAPI-3300I).

2.4.4 *Materials*

N, N-Dimethylformamide (DMF) was dried over the molecular sieve 4A and distilled under a decreased pressure (10–20 mmHg). TEOA was distilled under a decreased pressure (<1 mmHg). The DMF and TEOA were kept under Ar until they were used. BNAH [32], 4-methyl-4′-vinyl-2–2′-bipyridine (vbpy) [33], 4,4′-divinyl-2–2′-bipyridine (dvbpy) [34], Ru(N^N)$_2$Cl$_2$ (N^N = bpy or dmb) [35], **[Ru(C = C,bpy)]$^{2+}$** [33], **[Ru(C = C,dmb)]$^{2+}$** [33], **[Ir(C = C,ppy)]$^{+}$** [36], and 4′-bromo-2,2′-bipyridine [37–39], Re(CO)$_5$Br [40] were prepared according to the reported methods with some modifications. All of the other reagents were of reagent grade quality and were used without further purification.

2.4.5 *Synthesis*

2.4.5.1 Re(I) Complexes

Re(PEt$_3$)$_2$(CO)$_3$Br: Re(CO)$_5$Br (457.9 mg, 1.127 mmol) and PEt$_3$ ethanol solution (20wt%, 1.63 g, 2.759 mmol) were dissolved in toluene (60 mL, degassed using Ar) and refluxed under Ar for 12 days. Upon cooling to room temperature, the solvent was evaporated and brown oil was obtained. The oily products were separated by flash column chromatography on silica gel (eluent: toluene, first fraction). After removing the solvent under a decreased pressure, white solids were obtained, which were recrystallized with MeOH/water, washed with water, and then dried under vacuum.

Yield: 203.7 mg (31%). FT-IR (in CH_2Cl_2) ν_{CO}/cm^{-1}: 2044, 1940, 1888. ^{1}H NMR (297.60 MHz, acetone-d_6): δ/ppm, 2.12–2.02 (m, 12H, P($\underline{CH_2}CH_3$)$_3$), 1.16–1.06 (m, 18H, P($\underline{CH_2}CH_3$)$_3$).

[Re(Brbpy)(CO)$_2$(PEt$_3$)$_2$](PF$_6$) (**[Re(Br,Et)]**(PF$_6$)): Re(PEt$_3$)$_2$(CO)$_3$Br (100.0 mg, 0.1705 mmol) and silver triflate (48.5 mg, 0.189 mmol) were dissolved in THF (8 mL) and refluxed under Ar for 3 h. Precipitated AgBr was filtered with Celite. A yellow-green oil was collected after the solvent was removed under vacuum. This oil and 4-bromo-2,2′-bipyridine (Brbpy, 50.2 mg, 0.212 mmol) were dissolved in toluene (15 mL) and heated at 90 °C under Ar overnight. Brbpy (39.2 mg, 0.166 mmol) was added to the reaction mixture and heated overnight. After the second addition of Brbpy (55.6 mg, 0.235 mmol), the solution was further heated for 1 d. The solvent was evaporated to give an orange-red residue, which was purified by flash column chromatography on silica gel (collected as the second fraction using 100:0–100:5 (v/v) CH_2Cl_2/MeOH as the eluent). The obtained orange-red solid was dissolved in MeOH, to which a saturated methanol solution of NH_4PF_6 and a small amount of water were added. The solution was gradually evaporated until the orange-red solid precipitated. The solid was washed with water and Et_2O and dried at 60 °C under vacuum. Yield: 112.8 mg (77%, over two steps). ESI–MS (in MeCN) *m/z*: 713 ([M-PF$_6$]$^+$). FT-IR (in CH_2Cl_2) ν_{CO}/cm^{-1}: 1934, 1864. ^{1}H NMR (297.60 MHz, acetone-d_6): δ/ppm, 9.32 (d, J = 5.4 Hz, 1H, Brbpy-6), 9.14 (d, J = 6.0 Hz, 2H, Brbpy-6′), 9.02 (d, J = 2.2 Hz, 1H, Brbpy-3′), 8.90 (dd, J = 1.4, 8.2 Hz, 1H, Brbpy-3), 8.38 (dd, J = 6.9, 8.2 Hz, 1H, Brbpy-4), 8.03 (dd, J = 2.2, 6.0 Hz, 1H, Brbpy-5′), 7.89 (ddd, J = 1.4, 5.4, 6.9 Hz, 1H, Brbpy-5), 1.59–1.52 (m, 12H, P($\underline{CH_2}CH_3$)$_3$), 0.92–0.83 (m, 18H, P($CH_2\underline{CH_3}$)$_3$). Elemental Anal. Calcd. for $C_{24}H_{37}BrF_6N_2O_2P_3Re$: C, 33.57; H, 4.34; N, 3.26. Found: C, 33.94; H, 4.18; N, 3.30.

Re(P(OEt)$_3$)$_2$(CO)$_3$Br: Re(CO)$_5$Br (1.17 g, 2.89 mmol) and P(OEt)$_3$ (1.13 g, 6.82 mmol) were dissolved in toluene (100 mL), and the solution was refluxed under Ar for 28 h. After the solvent was evaporated, the residue was separated by column chromatography on alumina (collected as the first fraction, using 100:0–0:100 (v/v) CH_2Cl_2/Hexane as the eluents). The obtained white solids were dried under vacuum. Yield: 1.37 g (70%). FT-IR (in CH_2Cl_2) ν_{CO}/cm^{-1}: 2069, 2048, 1970, 1920. ^{1}H NMR (297.60 MHz, acetone-d_6): δ/ppm, 4.19–4.10 (m, 12H, P(O$\underline{CH_2}CH_3$)$_3$), 1.31 (t, J = 6.9 Hz, 18H, P(O$\underline{CH_2}CH_3$)$_3$).

[Re(vbpy)(CO)$_2${P(OEt)$_3$}$_2$](PF$_6$) (**[Re(C = C,OEt)]**(PF$_6$)): Re(P(OEt)$_3$)$_2$(CO)$_3$Br (210.0 mg, 0.3077 mmol) and silver triflate (68.65 mg, 0.3433 mmol) was dissolved in THF (20 mL), and the solution was refluxed under Ar for 4 h. Precipitated AgBr was filtered by Celite. Brown oil was collected after the solvent was evaporated. A toluene (25 mL) containing the oil and 4-methyl-4′-vinyl-2,2′-bipyridine (vbpy, 72.3 mg, 0.368 mmol) was heated at 90 °C under Ar for 21.5 h. Evaporation of the solvent gave a brown residue, which was separated by flash column chromatography on silica gel (collected as first yellow fraction, using 100:0–100:2 (v/v) CH_2Cl_2/MeOH as the eluent). The purification method was same as that for **[Re(Br,Et)]$^+$**. Yield: 145.8 mg (52%). ESI–MS (in MeCN) *m/z*: 771 ([M-PF$_6$]$^+$). FT-IR (in CH_2Cl_2) ν_{CO}/cm^{-1}: 1954, 1879. ^{1}H NMR

(297.60 MHz,acetone-d_6): δ/ppm, 8.95 (d, J = 5.7 Hz, 1H, vbpy-6′), 8.87 (d, J = 5.9 Hz, 1H, vbpy-6), 8.78 (s, 1H, vbpy-3′), 8.68 (s, 1H, vbpy-3), 7.80 (d, J = 5.9 Hz, 1H, vbpy-5), 7.61 (d, J = 5.7 Hz, 1H, vbpy-5′), 6.97 (dd, J = 17.6, 10.9 Hz, 1H, vbpy–C$\underline{H}$=CH$_2$), 6.44 (d, J = 17.6 Hz, 1H, vbpy–CH=C$\underline{H_2}$ (trans)), 5.76 (d, J = 10.9 Hz, 1H, vbpy–CH=C$\underline{H_2}$ (cis)), 3.83 (m, 12H, P(O–C$\underline{H_2}$–CH$_3$)$_3$), 2.60 (s, 3H, vbpy–CH$_3$), 1.00 (t, J = 6.9 Hz, 18H, P(O–CH$_2$–C$\underline{H_3}$)$_3$). Elemental Anal. Calcd for $C_{27}H_{42}F_6N_2O_8P_3Re$: C, 35.41; H, 4.62; N, 3.06. Found: C, 35.64; H, 4.40; N, 3.18.

Re(PPh$_3$)$_2$(CO)$_3$Br: A toluene solution (50 mL) containing Re(CO)$_5$Br (303 mg, 0.747 mmol) and PPh$_3$ (457 mg, 1.74 mmol) was refluxed under N_2 for 2 days. After cooling to room temperature, the white precipitation was collected and washed with toluene and hexane. Yield: 435 mg (67%). FT-IR (in CH_2Cl_2) ν_{CO}/cm^{-1}: 2051, 1948, 1910. Elemental Anal. Calcd. for $C_{39}H_{30}BrO_3P_2Re$: C, 53.55; H, 3.46. Found: C, 53.67; H, 3.66.

[Re(Br$_2$bpy)(CO)$_2$(PPh$_3$)$_2$](PF$_6$) (**[Re(Br$_2$,Ph)]**(PF$_6$)): Re(PPh$_3$)$_2$(CO)$_3$Br (298.6 mg, 0.3414 mmol) and silver triflate (105.0 mg, 0.4087 mmol) were dissolved in CH_2Cl_2 (30 mL) and the solution was refluxed under N_2 for 17 h. Precipitated AgBr was filtered by Celite. After evaporating the solvent, the obtained white residue and 4,4′-dibromo-2,2′-bipyridine (Br$_2$bpy, 129.4 mg, 0.4120 mmol) was dissolved in toluene (30 mL, degassed using N_2), and the solution was heated at 90 °C under N_2 for 16 h. Orange precipitation was filtered off, washed with toluene and hexane, and purified by flash column chromatography on silica gel (collected as first orange fraction, using 100:0–100:5 (v/v) CH_2Cl_2/MeOH as the eluents). The purification method was same as that for **[Re(Br,Et)]**$^+$. Yield: 321.7 mg (77%). ESI–MS (in MeCN) m/z: 1081 ([M-PF$_6$]$^+$). FT-IR (in CH_2Cl_2) ν_{CO}/cm^{-1}: 1942, 1874. ^{1}H NMR (297.60 MHz, acetone-d_6): δ/ppm, 8.15 (d, J = 1.8 Hz, 2H, bpy-3,3′), 7.65 (d, J = 6.0 Hz, 2H, bpy-6,6′), 7.16–7.30 (m, 30H, 2P$\underline{Ph_3}$), 6.84 (dd, J = 1.8, 6.0 Hz, 2H, bpy-5,5′). Elemental Anal. Calcd. for $C_{48}H_{36}Br_2F_6N_2O_2P_3Re$: C, 47.03; H, 2.96; N, 2.29. Found: C, 46.95; H, 3.03; N, 2.15.

[Re(dvbpy)(CO)$_2$(PPh$_3$)$_2$](PF$_6$) (**[Re(C = C$_2$,Ph)]**(PF$_6$)) was synthesized by a similar method for synthesis of **[Re(Br$_2$,Ph)]**$^+$ using dvbpy instead of Br$_2$bpy: Yield: 56%. ESI–MS (in MeCN) m/z: 975 ([M-PF$_6$]$^+$). FT-IR (in CH_2Cl_2) ν_{CO}/cm^{-1}: 1937, 1867. ^{1}H NMR (297.60 MHz, acetone-d_6): δ/ppm, 8.49 (d, J = 1.6 Hz, 2H, bpy-3,3′), 7.98 (d, J = 5.8 Hz, 2H, bpy-6,6′), 7.16–7.30 (m, 30H, 2P$\underline{Ph_3}$), 7.03 (dd, J = 1.6, 5.8 Hz, 2H, bpy-5,5′), 6.81 (dd, J = 10.6, 17.4 Hz, 2H, –C$\underline{H}$=CH$_2$), 6.27 (d, J = 10.6 Hz, 2H, –CH=C$\underline{H_2}$ (cis)), 5.73 (d, J = 17.4 Hz, 2H, –CH=C$\underline{H_2}$ (trans)). ^{31}P NMR (162 MHz, acetone-d_6): δ/ppm, 21.9 (s, 2P, $\underline{P}$Ph$_3$), – 143.6 (7, 1P, $\underline{P}$F$_6^-$). HRMS (ESI-TOF) m/z: [M-PF$_6^-$]$^+$ Calcd for $C_{52}H_{42}N_2O_2P_2Re$ 975.2283; Found 975.2284.

Re{P(p-FC$_6$H$_4$)$_3$}$_2$(CO)$_3$Br: Re(CO)$_5$Br (104 mg, 0.255 mmol) and P(p-FC$_6$H$_4$)$_3$ (165 mg, 0.522 mmol) were dissolved in toluene (20 mL), and the solution was refluxed under N_2 for 23 h. Evaporation of the solvent gave yellow oil, which was separated by column chromatography on silica gel (eluent: CH_2Cl_2).The obtained white solids were recrystallized with toluene-hexane, and dried under vacuum. Yield: 155 mg (62%). FT-IR (in CH_2Cl_2) ν_{CO}/cm^{-1}: 2057, 1955, 1911. ^{1}H NMR (399.65 MHz, acetone-d_6): δ/ppm, 7.69–7.75 (m, 18H, m-Ph), 7.26 (dd, J = 8.8,

8.8 Hz, 18H, *o*-Ph). Elemental Anal. Calcd for $C_{39}H_{24}BrF_6O_3P_2Re$: C, 47.67; H, 2.46. Found: C, 47.84; H, 2.70.

[Re(Br_2bpy)$(CO)_2${P(*p*-$FC_6H_4)_3$}$_2$](PF_6) (**[Re(Br_2,FPh)]**(PF_6)): Re{P(*p*-$FC_6H_4)_3$}$_2(CO)_3$Br (310 mg, 0.315 mmol) and silver triflate (92.5 mg, 0.359 mmol) was dissolved in THF (30 mL), and the solution was refluxed under Ar for 3 h. Precipitated AgBr was filtered by Celite. White solids were collected after the solvent was evaporated. The toluene solution (30 mL) containing the obtained residue and 4,4′-dibromo-2,2′-bipyridine (Br_2bpy, 119 mg, 0.380 mmol) was heated at 90 °C under N_2 for 16 h. After the solvent was evaporated, the obtained orange solids were purified by flash column chromatography on silica gel (collected as first orange fraction, using 100:0–100:5 (v/v) CH_2Cl_2/MeOH as the eluents). The purification method was same as that for **[Re(Br,Et)]$^+$**. Yield: 333 mg (79%). ESI–MS (in MeCN) *m/z*: 1189 ([Re(Br_2bpy)$(CO)_2${P(*p*-$FC_6H_4)_3$}$_2$]$^+$). FT-IR (in CH_2Cl_2) ν_{CO}/cm^{-1}: 1943, 1874. ^{1}H NMR (297.60 MHz, acetone-d_6): δ/ppm, 8.72 (d, J = 1.8 Hz, 2H, Br_2bpy-3,3′), 8.13 (d, J = 6.0 Hz, 2H, Br_2bpy-6,6′), 7.46–7.38 (m, 14H, bpy-5,5′, *m*-Ph), 7.14 (dd, J = 8.7, 8.7 Hz, 12 H, *o*-Ph). Elemental Anal. Calcd. for $C_{48}H_{36}Br_2F_6N_2O_2P_3Re$: C, 47.03; H, 2.96; N, 2.29. Found: C, 46.95; H, 3.03; N, 2.15.

[Re(dvbpy)$(CO)_2${P(*p*-$FC_6H_4)_3$}$_2$](PF_6) (**[Re(C = C_2,FPh)]**(PF_6)) was synthesized by a similar method for synthesis of **[Re(Br_2,FPh)]$^+$** using dvbpy instead of Br_2bpy: Yield: 51%. ESI–MS (in MeCN) *m/z*: 1083 ([M-PF_6]$^+$). FT-IR (in CH_2Cl_2) ν_{CO}/cm^{-1}: 1939, 1869. ^{1}H NMR (297.60 MHz, acetone-d_6): δ/ppm, 8.73 (d, J = 4.4 Hz, 1H, bpy-6), 8.62 (s, 1H, bpy-3), 8.53 (s, 1H, bpy-3′), 8.13 (d, J = 6.2 Hz, 1H, bpy-6′), 7.43–6.79 (m, 28H, 6 × P(*p*-$FC_6\underline{H_4})_3$, bpy-5,5′, 2 × –C$\underline{H}$=CH_2), 6.29 (d, J = 17.7 Hz, 2H, 2 × –CH=C$\underline{H_2}$(cis)), 5.75 (d, J = 11.2 Hz, 1H, –CH=C$\underline{H_2}$(trans)), 6.29 (d, J = 11.2 Hz, 1H, –CH=C$\underline{H_2}$(trans)). ^{31}P NMR (162 MHz, acetone-d_6): δ/ppm, 20.1 (s, 2P, $\underline{P}$(*p*-$FC_6H_4)_3$), − 143.6 (7, 1P, $\underline{P}F_6^-$). HRMS (ESI-TOF) *m/z*: [M-PF_6^-]$^+$ Calcd for $C_{52}H_{36}F_6N_2O_2P_2Re$ 1083.1718; Found: 1083.1689.

2.4.5.2 Synthesis of Multinuclear Complexes Using the Mizoroki–Heck Reaction

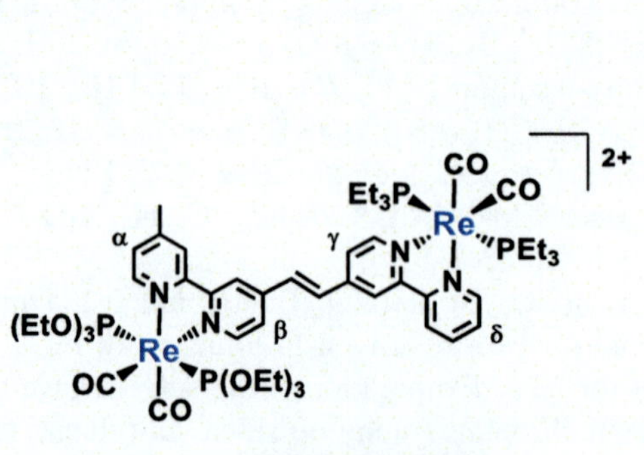

[Re(OEt)Re(Et)]$^{2+}$

[Re(OEt)Re(Et)](PF_6)$_2$: **[Re(Br,Et)]**(PF_6) (20 mg, 0.023 mmol), **[Re(C = C,OEt)]**(PF_6) (21 mg, 0.023 mmol), $Pd(OAc)_2$ (1.0 mg, 0.0045 mmol), PPh_3 (2.4 mg, 0.0092 mmol) and AcONa (11.5 mg, 0.140 mmol) were dissolved in MeCN degassed using Ar (4 mL). The solution was refluxed at 75 °C under Ar for 14 h. An MeCN solution (4 mL) containing $Pd(OAc)_2$ (1.0 mg, 0.0045 mmol) and PPh_3 (2.4 mg, 0.0092 mmol) was added to the reaction mixture and refluxed for 20 h. The solvent was removed under vacuum to give an orange-red solid. The coupling reagents were removed by flash column chromatography on silica gel (100:0–100:2 CH_2Cl_2/MeOH as the eluent). The target complex was isolated using ion-exchange chromatography [SP Sephadex C-25; the eluent was a 1:1 v/v mixture of MeCN and water containing NH_4PF_6 (0–4 mM)]. The solvent was evaporated until the solid precipitated. The solid was washed with water and Et_2O and dried at 60 °C under vacuum. Yield: 27.7 mg (70%). ^{1}H NMR (400 MHz, acetone-d_6): δ/ppm, 9.30 (d, J = 5.7 Hz, 1H, γpy-6), 9.28 (d, J = 5.8 Hz, 1H, βpy-6), 9.10 (d, J = 5.8 Hz, 1H, δpy-6), 9.01 (s, 1H, γpy-3), 8.95 (s, 1H, βpy-3), 8.94 (s, 1H, δpy-3), 8.87 (d, J = 7.1 Hz, αpy-6), 8.67 (s, 1H, αpy-3), 8.37 (dd, J = 7.1, 8.0 Hz, 1H, δpy-4), 8.07 (m, 2H, –C$\underline{H}$=C$\underline{H}$–), 8.03 (d, J = 5.8 Hz, 1H, βpy-5), 8.00 (d, J = 5.7 Hz, 1H, γpy-5), 7.87 (dd, J = 5.8, 7.1 Hz, 1H, δpy-5), 7.68 (d, J = 7.1 Hz, 1H, αpy-5), 3.92–3.86 (m, 12H, P(OC$\underline{H_2}$CH$_3$)$_3$), 2.67 (s, 3H, vbpy–C$\underline{H_3}$), 1.60–1.53 (m, 12H, P(C$\underline{H_2}$CH$_3$)$_3$), 1.04 (t, J = 7.3 Hz, 18H, P(OCH$_2$C$\underline{H_3}$)$_3$), 0.94–0.84 (m, 18H, P(CH$_2$C$\underline{H_3}$)$_3$). ^{31}P NMR (162 MHz, acetone-d_6): δ/ppm, 117.0 (s, 2P, $\underline{P}$(OEt)$_3$), – 2.06 (s, 2P, $\underline{P}$Et$_3$), – 143.5 (7, 2P, $\underline{P}F_6^-$). FT-IR (in CH_2Cl_2) ν_{CO}/cm^{-1}: 1954, 1930, 1880, 1860. ESI–MS (in MeCN) *m/z*: 702 ([M-2PF$_6^-$]$^{2+}$). Elemental Anal. Calcd for $C_{51}H_{78}F_{12}N_4O_{10}P_6Re_2$: C, 36.17; H, 4.64; N, 3.31. Found: C, 36.05; H, 4.30; N, 3.33. HRMS (ESI-TOF) *m/z*: [M-2PF$_6^-$]$^{2+}$ Calcd for $C_{51}H_{78}N_4O_{10}P_4Re_2$ 701.1879; Found: 701.1848.

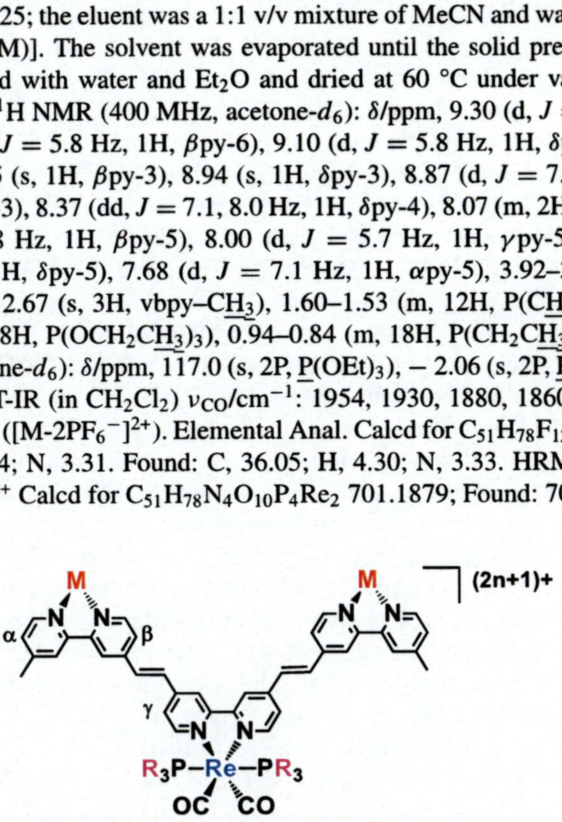

[Re(OEt)$_2$Re(Ph)]$^{3+}$: M = Re^{I}(CO)$_2$(P(OEt)$_3$)$_2$, R = Ph, n = 1
[Ru(bpy)$_2$Re(Ph)]$^{5+}$: M = Ru^{II}(bpy)$_2$, R = Ph, n = 2
[Ru(dmb)$_2$Re(Ph)]$^{5+}$: M = Ru^{II}(dmb)$_2$, R = Ph, n = 2
[Ir(ppy)$_2$Re(Ph)]$^{3+}$: M = Ir^{III}(ppy)$_2$, R = Ph, n = 1
[Ru(dmb)$_2$Re(FPh)]$^{5+}$: M = Ru^{II}(dmb)$_2$, R = P(p-FC$_6$H$_4$)$_3$, n = 2

[Re(OEt)$_2$Re(Ph)](PF_6)$_3$: **[Re(Br$_2$,Ph)]**(PF_6) (10 mg, 0.0081 mmol), **[Re(C = C,OEt)]**(PF_6) (15 mg, 0.016 mmol), $Pd(OAc)_2$ (0.37 mg, 0.0016 mmol), PPh_3 (0.85 mg, 0.0032 mmol) and AcONa (3.3 mg, 0.041 mmol) were dissolved in MeCN degassed by N_2 (2 mL). The solution was refluxed at around 75 °C under N_2 for

2 days in dim light. An MeCN solution (2 mL) containing $Pd(OAc)_2$ (0.37 mg, 0.0016 mmol) and PPh_3 (0.85 mg, 0.0032 mmol) were added and refluxed 2 days again. Evaporation of the solvent gave brown solids, which were twice purified by ion-exchange chromatography (SP Sephadex C-25 and CM Sephadex C-25; the eluent was a 1:1 v/v mixture of MeCN and water containing NH_4PF_6 (0–8 mM)). The solvent was evaporated until orange solids were precipitated. The solid was washed with water and Et_2O, and then dried at 60 °C under vacuum. Yield: 7.4 mg (31%). ^{1}H NMR (400 MHz, acetone-d_6): δ/ppm, 9.11 (d, $J = 5.8$ Hz, 2H, βpy-6), 8.97 (d, $J = 5.2$ Hz, 2H, αpy-6), 8.92 (s, 2H, βpy-3), 8.71 (s, 2H, αpy-3), 8.57 (s, 2H, γpy-3), 8.13 (d, $J = 5.2$ Hz, 2H, γpy-6), 8.03 (d, $J = 5.8$ Hz, 1H, βpy-5), 7.93 (d, $J = 5.2$ Hz, 2H, γpy-5), 7.72 (d, $J = 5.2$ Hz, 2H, αpy-5), 7.42–7.28 (m, 34H, –C<u>H</u>=C<u>H</u>–, 2P<u>Ph</u>$_3$), 3.95–3.88 (m, 24H, $4P(OC\underline{H_2}CH_3)_3$), 2.74 (s, 6H, αpy–$\underline{CH_3}$), 1.08 (t, $J = 7.1$ Hz, 36H, $4P(OCH_2\underline{CH_3})_3$). ^{31}P NMR (162 MHz, acetone-d_6): δ/ppm, 117.0 (s, 4P, –$\underline{P}(OEt)_3$), 21.8 (s, 2P, –$\underline{P}Ph_3$), – 143.4 (7, 3P, $\underline{P}F_6^-$). FT-IR (in CH_2Cl_2) ν_{CO}/cm^{-1}: 1954, 1940, 1878(br). ESI–MS (in MeCN) *m/z*: 820 ([M-$3PF_6^-]^{3+}$), 1303 ([M-$2PF_6^-]^{2+}$). HRMS (ESI-TOF) *m/z*: [M-$3PF_6^-]^{5+}$ Calcd for $C_{102}H_{118}N_6O_{18}P_6Re_3$ 819.8528; Found: 819.8522.

[Ru(bpy)$_2$Re(Ph)]$(PF_6)_5$: **[Re(Br$_2$,Ph)]**(PF_6) (15 mg, 0.012 mmol), **[Ru(C = C,bpy)]**$(PF_6)_2$ (25 mg, 0.028 mmol), $Pd(OAc)_2$ (0.52 mg, 0.0023 mmol), PPh_3 (1.5 mg, 0.0057 mmol) and AcONa (1.5 mg, 0.018 mmol) were dissolved in MeCN degassed using N_2 (10 mL). The solution was refluxed at 75 °C under N_2 for 17 h in dim light. An MeCN solution degassed using N_2 (10 mL) containing $Pd(OAc)_2$ (0.52 mg, 0.0023 mmol) and PPh_3 (1.3 mg, 0.0050 mmol) was added to the reaction mixture and refluxed for 1 d. $Pd(OAc)_2$ (1.0 mg, 0.0045 mmol) and PPh_3 (2.4 mg, 0.0092 mmol) were again added to the solution and the mixture was refluxed for 1 d. Evaporation of the solvent gave a red–black solid, which was purified by size exclusion chromatography. The red portion of the eluted solution was evaporated, dissolved in CH_2Cl_2 and washed twice with water containing NH_4PF_6. The red solid was obtained from the organic phase and dissolved in MeOH. A saturated methanol solution of NH_4PF_6 and a small amount of water were added to the dissolved solid. Evaporation of methanol gave a red solid, which was then washed with water and Et_2O and dried at 60 °C under vacuum. Yield: 15 mg (42%). ^{1}H NMR (400 MHz, acetone-d_6): δ/ppm, 8.98 (s, 2H), 8.85–8.83 (m, 8H), 8.76 (s, 2H), 8.51 (s, 2H), 8.23–8.20 (m, 10H), 8.12–8.07 (m, 10H), 7.89 (d, $J = 5.8$ Hz, 2H), 7.84 (s, 4H), 7.73 (d, $J = 5.8$ Hz, 8H), 7.63–7.59 (m, 8H), 7.45 (d, $J = 5.4$ Hz, 2H), 7.32 (m, 30H, 2P<u>Ph</u>$_3$), 7.22 (d, $J = 5.8$ Hz, 2H), 2.65 (s, 6H, αpy–$\underline{CH_3}$). ^{31}P NMR (162 MHz, acetone-d_6): δ/ppm, 21.8 (s, 2P, -$\underline{P}Ph_3$), – 143.5 (7, 5P, $\underline{P}F_6^-$). FT-IR (in CH_2Cl_2) ν_{CO}/cm^{-1}: 1938, 1870. ESI–MS (in MeCN) *m/z*: 426 ([M-$5PF_6^-]^{5+}$), 571 ([M-$4PF_6^-]^{4+}$), 810 ([M-$3PF_6^-$)]$^{3+}$). Elemental Anal. Calcd for $C_{114}H_{90}F_{30}N_{14}O_2P_7ReRu_2$: C, 47.82; H, 3.17; N, 6.85. Found C, 47.61; H, 3.22; N, 6.58. HRMS (ESI-TOF) *m/z*: [M-$5PF_6^-]^{5+}$ Calcd for $C_{114}H_{90}N_{14}O_2P_2ReRu_2$ 427.8907; Found: 427.8919, [M-$4PF_6^-]^{4+}$ Calcd for $C_{114}H_{90}F_6N_{14}O_2P_3ReRu_2$ 571.1044; Found: 571.1010, [M-$3PF_6^-]^{3+}$ Calcd for $C_{114}H_{90}F_{12}N_{14}O_2P_4ReRu_2$ 809.7939; Found: 809.7933.

[Ru(dmb)$_2$Re(Ph)]$(PF_6)_5$: **[Re(Br$_2$,Ph)]**(PF_6) (30 mg, 0.024 mmol), **[Ru(C = C,dmb)]**$(PF_6)_2$ (47 mg,0.049 mmol), $Pd(OAc)_2$ (5.5 mg, 0.024 mmol), PPh_3 (13 mg,

0.049 mmol) and AcONa (12 mg, 0.146 mmol) were dissolved in MeCN degassed using N_2 (4 mL). The solution was refluxed at around 75 °C under N_2 for 1 d in dim light. An MeCN solution degassed using N_2 (4 mL) containing $Pd(OAc)_2$ (5.5 mg, 0.024 mmol) and PPh_3 (13 mg, 0.049 mmol) was added and then the solution was refluxed for 1 d. The solvent was removed under vacuum giving red-black solids. The solids were separated by size exclusion chromatography, and the red portion of the eluted solution was evaporated. The residue was dissolved in CH_2Cl_2 and twice washed with water containing NH_4PF_6. After evaporation, the obtained red solids were dissolved in MeOH, and a saturated methanol solution of NH_4PF_6 and a small amount of water were added to the solution. Evaporation of MeOH gave red solids, which were washed with water and Et_2O, and dried at 60 °C under vacuum. Yield: 23 mg (32%).^{1}H NMR (300 MHz, acetone-d_6): δ/ppm, 8.96 (s, 2H, βpy-3), 8.74 (s, 2H, αpy-3), 8.69 (s, 8H, dmb-3,3′), 8.51 (s, 2H, γpy-3), 8.10 (d, $J = 6.2$ Hz, 2H, γpy-6), 8.08 (d, $J = 7.3$ Hz, 2H, βpy-6), 7.98 (d, $J = 5.8$ Hz, 2H, γpy-5), 7.85–7.81 (m, 10H, αpy-6, dmb-6,6′), 7.72 (d, $J = 5.6$ Hz, 2H, βpy-5), 7.45–7.22 (m, 40H, -6P$\underline{Ph}_3$, dmb-5,5′, –$\underline{CH}$=$\underline{CH}$–), 7.23 (d, $J = 6.2$ Hz, 2H, αpy-5), 2.65 (s, 6H, αpy–$C\underline{H}_3$), 2.56 (s, 24H, dmb–$\underline{CH_3}$). ^{31}P NMR (162 MHz, acetone-d_6): δ/ppm, 21.7 (s, 2P, -$\underline{P}Ph_3$), – 143.5 (7, 5P, $\underline{P}F_6^-$). FT-IR (in CH_2Cl_2) ν_{CO}/cm^{-1}: 1939, 1870. ESI–MS (in MeCN) *m/z*: 450 ([M-5PF_6^-]$^{5+}$), 599 ([M-4PF_6^-]$^{4+}$), 847 ([M-3PF_6^-]$^{3+}$). HRMS (ESI-TOF) *m/z*: [M-5PF_6^-]$^{5+}$ Calcd for $C_{122}H_{106}N_{14}O_2P_2ReRu_2$ 450.3158; Found: 450.3158, [M-4PF_6^-]$^{4+}$ Calcd for $C_{122}H_{106}F_6N_{14}O_2P_3ReRu_2$ 599.1357; Found: 599.1362, [M-3PF_6^-]$^{3+}$ Calcd for $C_{122}H_{106}F_{12}N_{14}O_2P_4ReRu_2$ 847.1691; Found: 847.1655.

[Ir(ppy)$_2$Re(Ph)](PF_6)$_3$: **[Re(Br$_2$,Ph)]**(PF_6) (11 mg, 9.2 μmol), **[Ir(C = C,ppy)]**(PF_6) (18 mg, 22 μmol), $Pd(OAc)_2$ (1.1 mg, 4.9 μmol), PPh_3 (2.8 mg, 11 μmol), and AcONa (1.6 mg, 19 μmol) were dissolved in MeCN degassed using N_2 (10 mL). The solution was refluxed at around 75 °C under N_2 for 1 d in dim light. A MeCN solution degassed using N_2 (10 mL) containing $Pd(OAc)_2$ (0.64 mg, 2.9 μmol) and PPh_3 (1.2 mg, 4.6 μmol) was added and the reaction solution was refluxed for 1 d. A MeCN solution degassed using N_2 (10 mL) containing $Pd(OAc)_2$ (0.50 mg, 2.2 μmol) and PPh_3 (1.2 mg, 4.6 μmol) was added again and the solution was refluxed for 1 d. Evaporation of the solvent gave brown solids, which were separated by size exclusion chromatography. The orange-brown portion of the eluted solution was evaporated, and the residue was dissolved in CH_2Cl_2 and twice washed with water containing NH_4PF_6. After evaporation of the organic phase, the orange solids were dissolved in methanol, and a saturated methanol solution of NH_4PF_6 and a small amount of water were added to the solution. After evaporation of methanol, obtained orange solids were collected, washed with water and Et_2O, and dried at 60 °C under vacuum. Yield: 2.0 mg (8.0%). ^{1}H NMR (400 MHz, acetone-d_6): δ/ppm, 9.01 (s, 2H, βpy-3), 8.78 (s, 2H, αpy-3), 8.56 (s, 2H, γpy-3), 8.25 (m, 6H), 8.09 (dd, $J = 5.7$, 6.2 Hz, 2H), 7.97–7.84 (m, 8H), 7.57 (d, $J = 6.0$ Hz, 2H), 7.32–7.16 (m, 38H), 7.07–6.93 (m, 10H), 6.36 (dd, $J = 7.3$, 7.4 Hz, 2H), 2.68 (s, 6H, αpy–$\underline{CH_3}$). ^{31}P NMR (162 MHz, acetone-d_6): δ/ppm, 21.8 (s, 2P, -$\underline{P}Ph_3$), – 143.5 (7, 3P, $\underline{P}F_6^-$). FT-IR (in CH_2Cl_2) ν_{CO}/cm^{-1}: 1939, 1870. ESI–MS (in MeCN) *m/z*: 771 ([M-3PF_6^-]$^{3+}$), 1229 ([M-2PF_6^-]$^{4+}$). HRMS (ESI-TOF) *m/z*: [M-3PF_6^-]$^{5+}$ Calcd for $C_{118}H_{90}Ir_2N_{10}O_2P_2Re$ 770.8518; Found: 770.8478.

[Ru(dmb)$_2$Re(FPh)](PF_6)$_5$ was synthesized from **[Ru(C = C,dmb)]$^{2+}$** and by a similar method to **[Ru(dmb)$_2$Re(Ph)]$^{5+}$** using **[Re(Br$_2$,FPh)]$^+$** instead of **[Re(Br$_2$,Ph)]$^+$**. Yield: 25%. ^{1}H NMR (400 MHz, acetone-d_6): δ/ppm, 8.94 (s, 2H, βpy-3′), 8.71 (s, 2H, αpy-3), 8.66 (s, 8H, dmb-3,3′), 8.56 (s, 2H, γpy-3), 8.18 (d, J = 6.0 Hz, 2H, γpy-6), 8.08 (d, J = 6.0 Hz, 2H, βpy-6′), 7.96 (d, J = 5.6 Hz, 2H, γpy-5), 7.86–7.82 (m, 12H, dmb-6,6′, –C<u>H</u>=C<u>H</u>–), 7.69 (d, J = 5.6 Hz, 2H, αpy-6), 7.46–7.38 (m, 24H, αpy-5, βpy-5′, dmb-5,5′, *m*-Ph), 7.13–7.09 (m, 12 H, *o*-Ph), 2.63 (s, 3H, αpy–C<u>H</u>$_3$), 2.56 (s, 24H, dmb–C<u>H</u>$_3$). ^{31}P NMR (162 MHz, acetone-d_6): δ/ppm, 19.9 (s, 2P, -<u>P</u>Ph$_3$), – 143.6 (7, 5P, <u>P</u>F$_6^-$). FT-IR (in CH_2Cl_2) ν_{CO}/cm^{-1}: 1942, 1873. ESI–MS (in MeCN) *m/z*: 472 ([M-5PF$_6^-$]$^{5+}$), 626 ([M-4PF$_6^-$]$^{4+}$), 883 ([M-3PF$_6^-$]$^{3+}$). HRMS (ESI-TOF) *m/z*: [M-5PF$_6^-$]$^{5+}$ Calcd for $C_{122}H_{100}F_6N_{14}O_2P_2ReRu_2$ 471.9047; Found: 471.9044, [M-4PF$_6^-$]$^{4+}$ Calcd for $C_{122}H_{100}F_{12}N_{14}O_2P_3ReRu_2$ 626.1226; Found: 626.1216, [M-3PF$_6^-$]$^{3+}$ Calcd for $C_{122}H_{100}F_{18}N_{14}O_2P_4ReRu_2$ 883.1502; Found: 883.1474.

Acknowledgements This chapter was reproduced with some modification from the following reference by permission of The Royal Society of Chemistry:

<u>Y. Yamazaki</u>, T. Morimoto, O. Ishitani, "Synthesis of Novel Photofunctional Multinuclear Complexes Using a Coupling Reaction", *Dalton Trans.*, **2015**, *44*, 11626–11635.

References

1. Kaveevivitchai N, Chitta R, Zong R, El Ojaimi M, Thummel RP (2012) A molecular light-driven water oxidation catalyst. J Am Chem Soc 134:10721
2. Li F, Jiang Y, Zhang B, Huang F, Gao Y, Sun L (2012) Towards a solar fuel device: light-driven water oxidation catalyzed by a supramolecular assembly. Angew Chem Int Ed 51:2417
3. Ozawa H, Sakai K (2011) Photo-hydrogen-evolving molecular devices driving visible-light-induced water reduction into molecular hydrogen: structure-activity relationship and reaction mechanism. Chem Commun 47:2227
4. Arachchige SM, Brown JR, Chang E, Jain A, Zigler DF, Rangan K, Brewer KJ (2009) Design considerations for a system for photocatalytic hydrogen production from water employing mixed-metal photochemical molecular devices for photoinitiated electron collection. Inorg Chem 48:1989
5. Stoll T, Gennari M, Fortage J, Castillo CE, Rebarz M, Sliwa M, Poizat O, Odobel F, Deronzier A, Collomb M-N (2014) An efficient Ru^{II}–Rh^{III}–Ru^{II} polypyridyl photocatalyst for visible-light-driven hydrogen production in aqueous solution. Angew Chem Int Ed 53:1654
6. Koike K, Naito S, Sato S, Tamaki Y, Ishitani O (2009) Architecture of supramolecular metal complexes for photocatalytic CO_2 reduction III: effects of length of alkyl chain connecting photosensitizer to catalyst. J Photochem Photobiol A 207:109
7. Tamaki Y, Morimoto T, Koike K, Ishitani O (2012) Photocatalytic CO_2 reduction with high turnover frequency and selectivity of formic acid formation using Ru(II) multinuclear complexes. Proc Natl Acad Sci U S A 109:15673
8. Tamaki Y, Watanabe K, Koike K, Inoue H, Morimoto T, Ishitani O (2012) Development of highly efficient supramolecular CO_2 reduction photocatalysts with high turnover frequency and durability. Faraday Discuss 155:115
9. Tamaki Y, Koike K, Morimoto T, Ishitani O (2013) Substantial improvement in the efficiency and durability of a photocatalyst for carbon dioxide reduction using a benzoimidazole derivative as an electron donor. J Catal 304:22

10. Gholamkhass B, Mametsuka H, Koike K, Tanabe T, Furue M, Ishitani O (2005) Architecture of supramolecular metal complexes for photocatalytic CO_2 reduction: Ruthenium-rhenium bi- and tetranuclear complexes. Inorg Chem 44:2326
11. Kato E, Takeda H, Koike K, Ohkubo K, Ishitani O (2015) Ru(II)–Re(I) binuclear photocatalysts connected by $-CH_2XCH_2-$ (X = O, S, CH_2) for CO_2 reduction. Chem Sci 6:3003
12. Hwang I-W, Ko DM, Ahn TK, Yoon ZS, Kim D, Peng X, Aratani N, Osuka A (2005) Excitation energy migration in a dodecameric porphyrin wheel. J Phys Chem B 109:8643
13. Balzani V, Campagna S, Denti G, Juris A, Serroni S, Venturi M (1998) Designing dendrimers based on transition-metal complexes. Light-harvesting properties and predetermined redox patterns. Acc Chem Res 31:26
14. Connors PJ, Tzalis D, Dunnick AL, Tor Y (1998) Coordination compounds as building blocks: single-step synthesis of heteronuclear multimetallic complexes containing Ru^{II} and Os^{II}. Inorg Chem 37:1121
15. Tzalis D, Tor Y (1996) Coordination compounds as building blocks: single-step synthesis of multi-ruthenium(II) complexes. Chem Commun 1043
16. Goeb S, De Nicola A, Ziessel R (2005) Controlled synthesis of multinuclear metal complex arrays by cross coupling of coordinated ligands. J Org Chem 70:6802
17. Haak RM, Martinez Belmonte M, Escudero-Adan EC, Benet-Buchholz J, Kleij AW (2010) Olefin metathesis as a tool for multinuclear Co(iii)salen catalyst construction: access to cooperative catalysts. Dalton Trans 39:593
18. Haak RM, Castilla AM, Martinez Belmonte M, Escudero-Adan EC, Benet-Buchholz J, Kleij AW (2011) Access to multinuclear salen complexes using olefin metathesis. Dalton Trans 40:3352
19. Cassidy L, Horn S, Cleary L, Halpin Y, Browne WR, Vos JG (2009) Synthesis of asymmetric supramolecular compounds using a Ni(0) catalysed homo-coupling approach. Dalton Trans 3923
20. Arm KJ, Williams JAG (2006) A cross-coupling strategy for the synthesis of dimetallic assemblies containing mixed bipyridine-terpyridine bridging ligands: luminescence and energy transfer properties. Dalton Trans 2172
21. Whittle VL, Williams JAG (2009) Cyclometallated, bis-terdentate iridium complexes as linearly expandable cores for the construction of multimetallic assemblies. Dalton Trans 3929
22. Welter S, Salluce N, Belser P, Groeneveld M, De Cola L (2005) Photoinduced electronic energy transfer in modular, conjugated, dinuclear Ru(II)/Os(II) complexes. Coord Chem Rev 249:1360
23. Beletskaya IP, Cheprakov AV (2000) The heck reaction as a sharpening stone of palladium catalysis. Chem Rev 100:3009
24. Heck RF, Nolley JP (1972) Palladium-catalyzed vinylic hydrogen substitution reactions with aryl, benzyl, and styryl halides. J Org Chem 37:2320
25. Rigorously excluding air contamination in the reaction vessel dramatically decreased the yield of **[Re(OEt)Re(Et)]$^{2+}$**. However, adding too much air to the reaction vessel also decreased the yield. Air contamination that occurred when the cap of the vessel was opened to allow the reagents to be added supplied a suitable amount of air for **[Re(OEt)Re(Et)]$^{2+}$** to be produced. The role of air in the reaction is described in detail in Chapter 3.2.1
26. Takeda H, Yamamoto Y, Nishiura C, Ishitani O (2006) Analysis and isolation of cationic rhenium(I) and ruthenium(II) multinuclear complexes using size-exclusion chromatography. Anal Sci 22:545
27. Strouse GF, Schoonover JR, Duesing R, Boyde S, Jones WE, Meyer TJ (1995) Influence of electronic delocalization in metal-to-ligand charge transfer excited states. Inorg Chem 34:473
28. Bian Z-Y, Wang H, Fu W-F, Li L, Ding A-Z (2012) Two bifunctional Ru^{II}/Re^{I} photocatalysts for CO_2 reduction: A spectroscopic, photocatalytic, and computational study. Polyhedron 32:78
29. Liu S, Schanze KS (2004) Solvent tuned excited state configuration mixing in a π-conjugated metal–organic oligomer. Chem Commun 1510
30. Allen GH, White RP, Rillema DP, Meyer TJ (1984) Synthetic control of excited-state properties. Tris-chelate complexes containing the ligands 2,2'-Bipyrazine, 2,2'-Bipyridine, and 2,2'-Bipyrimidine. J Am Chem Soc 106:2613

31. Koike K, Okoshi N, Hori H, Takeuchi K, Ishitani O, Tsubaki H, Clark IP, George MW, Johnson FPA, Turner JJ (2002) Mechanism of the photochemical ligand substitution reactions of *fac*-$[Re(bpy)(CO)_3(PR_3)]^+$ complexes and the properties of their triplet ligand-field excited states. J Am Chem Soc 124:11448
32. Mauzerall D, Westheimer FH (1955) 1-Benzyldihydronicotinamide—a Model for Reduced DPN. J Am Chem Soc 77:2261
33. Guarr TF, Anson FC (1987) Electropolymerization of ruthenium (bis(1,10-phenanthroline)(4-methyl-4'-vinyl2,2'-bipyridine) complexes through direct attack on the ligand ring system. J Phys Chem 91:4037
34. Font J, de March P, Busque F, Casas E, Benitez M, Teruel L, Garcia H (2007) Periodic mesoporous silica having covalently attached tris(bipyridine)ruthenium complex: synthesis, photovoltaic and electrochemiluminescent properties. J Mater Chem 17:2336
35. Sullivan BP, Salmon DJ, Meyer TJ (1978) Mixed phosphine 2,2'-bipyridine complexes of ruthenium. Inorg Chem 17:3334
36. Metz S, Bernhard S (2010) Robust photocatalytic water reduction with cyclometalated Ir(III) 4-vinyl-2,2'-bipyridine complexes. Chem Commun 46:7551
37. Baron A, Herrero C, Quaranta A, Charlot MF, Leibl W, Vauzeilles B, Aukauloo A (2012) Click chemistry on a ruthenium polypyridine complex. An efficient and versatile synthetic route for the synthesis of photoactive modular assemblies. Inorg Chem 51:5985
38. Hasegawa E, Seida T, Chiba N, Takahashi T, Ikeda H (2005) Contrastive photoreduction pathways of benzophenones governed by regiospecific deprotonation of imidazoline radical cations and additive effects. J Org Chem 70:9632
39. Zhu X-Q, Zhang M-T, Yu A, Wang C-H, Cheng J-P (2008) Hydride, hydrogen atom, proton, and electron transfer driving forces of various five-membered heterocyclic organic hydrides and their reaction intermediates in acetonitrile. J Am Chem Soc 130:2501
40. Kutal C, Weber MA, Ferraudi G, Geiger D (1985) A mechanistic investigation of the photoinduced reduction of carbon dioxide mediated by tricarbonylbromo(2,2'-bipyridine)rhenium(I). Organometallics 4:2161

Chapter 3
Selective Synthesis of Various Photofunctional Multinuclear Complexes Using a Combination of the Mizoroki–Heck Reaction and the Homo-coupling Reaction

Abstract As described in Chap. 2, the Mizoroki–Heck reaction between metal complexes with bromo groups and those with vinyl groups in the diimine ligand did not proceed using $Pd(OAc)_2$ and a tertiary phosphine ligand under an Ar atmosphere. However, it proceeded selectively after the injection of air into the reaction vessel. In this chapter, I report the effect of the air injection on the Pd compounds for the coupling reactions in detail. Interestingly, the Mizoroki–Heck reaction took place even in the absence of the phosphine ligand. However, a homo-coupling reaction between the metal complexes with bromo groups proceeded at the same time. Mechanistic investigation showed that nanoparticles of Pd species were produced in the absence of the phosphine ligand and functioned as catalysts for both the Mizoroki–Heck and homo-coupling reactions. In contrast, larger Pd particles, which were produced in the presence of the phosphine ligands but after the addition of air for oxidizing the phosphine ligands, catalyzed the Mizoroki–Heck reaction with high selectivity. Based on these results, reaction conditions were established for the selective progress of the Mizoroki–Heck and the homo-coupling reactions.

Keywords Mizoroki–Heck reaction · Homo-coupling reaction · Pd-particles catalysts

3.1 Introduction

In Chap. 2, I reported that the Mizoroki–Heck reaction [1, 2] could be applied for coupling various photofunctional metal complexes [3]. The reaction proceeded after adding reagents for the coupling reaction with exposure to the ambient air. This result might indicate that some air is required to promote the coupling reaction. Therefore, investigation of the role of air in the coupling reaction should be important in order to consider suitable reaction conditions.

In this chapter, I will report the role of air in the coupling reaction and the concentration effect of the phosphine ligands on the reaction. From these results, it was indicated that phosphine ligands were not required for the coupling reaction. Rather, they could suppress the coupling reaction under anaerobic conditions. However, in the absence of phosphine ligands, some byproducts derived from side reaction,

Y. Yamazaki, *Development of Synthetic Methods for Novel Photofunctional Multinuclear Complexes*, Springer Theses, https://doi.org/10.1007/978-981-16-7148-7_3

i.e., homo-coupling [4–6], were observed, even though the Mizoroki–Heck reaction proceeded simultaneously. The existence of phosphine ligands in the initial stage should improve the reaction selectivity for the Mizoroki–Heck reaction, even though the phosphine ligands should be oxidized and removed from Pd compounds to start the reaction. When using no phosphine ligands, reaction selectivity for the Mizoroki–Heck reaction was improved by using an excess amount of complexes with vinyl groups. Therefore, the selective Mizoroki–Heck reaction should be achieved with two methods: (1) by using an excess amount of the complexes with vinyl groups in the absence of phosphine ligands, or (2) by introducing some air during heating in the presence of phosphine ligands. Interestingly, the homo-coupling reaction could also proceed selectively when using only complexes possessing bromo groups. Novel multinuclear metal complexes, i.e., a trinuclear complex having three different kinds of complex units and a tetranuclear complex having two kinds of diimine moieties, were successfully synthesized by a combination of the Mizoroki–Heck reactions and/or the homo-coupling reaction.

In this chapter, different abbreviations from those in Chap. 2 are used for simplicity.

3.2 Results and Discussion

3.2.1 The Effect of Air on the Mizoroki–Heck Reaction

In Chap. 2, I reported the synthesis of hetero-trinuclear metal complexes, such as **Ru$_2$Re**, from mononuclear complexes having a vinyl group or bromo groups as building blocks by using the Mizoroki–Heck reaction (Scheme 3.1). Although the coupling reaction did not undergo well regardless of amounts of the reagents for the coupling reaction, i.e., Pd(OAc)$_2$ and phosphine ligands [PPh$_3$ or 2-dicyclohexylphosphino-2′, 6’-dimethoxybiphenyl (Sphos)], under anaerobic conditions, the reaction proceeded after the reagents were added to the solution as an

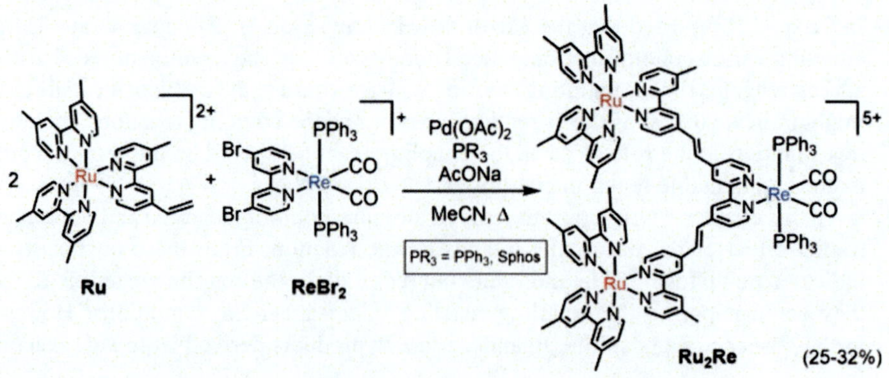

Scheme 3.1 A synthetic scheme of **Ru$_2$Re** by using the Mizoroki–Heck reaction

additional charge with exposure to ambient air. Therefore, the effect of air on the coupling reactions should be investigated in detail.

The blue line in Fig. 3.1 illustrates the size exclusion chromatogram [7] measured after heating the reaction solution at 65 °C under Ar for 2 days. The peaks attributed to **Ru** and **ReBr$_2$** were mainly observed, which indicated that the Mizoroki–Heck reaction did not proceed well. After injecting some air by using a syringe through the septum without additional charge of the reagents followed by further heating for 2 days, an obvious peak derived from **Ru$_2$Re** appeared in the chromatogram (Fig. 3.1 red line). This result indicates that not charging the reagents but injecting some air was necessary to promote the reaction. Since some drops of water instead of charging the reagents and/or some air could not promote the reaction, not moisture but some oxygen should play an important role in the reaction.

^{31}P NMR measurement of the reaction solution after injecting some air showed that all of the phosphine ligands were converted to oxidized species (the oxidized Sphos: $\delta = 47.6$ [ppm]). Since the reaction continuously proceeded after oxidation of the phosphine ligands, Pd compounds having no phosphine ligand should function as catalysts for the Mizoroki–Heck reaction.

The reaction without phosphine ligands and the additional charge of Pd(OAc)$_2$ under an Ar atmosphere was performed. **Ru** (8 μmol), **ReBr$_2$** (4 μmol), Pd(OAc)$_2$ (4 μmol), and AcONa (21 μmol) were dissolved in MeCN (1 mL) and the solution was heated at 65 °C for 3 days under Ar. After the reaction, the crude products were analyzed by size exclusion chromatography (SEC, Fig. 3.2 red line). The main peak detected at around 33 min was attributed to **Ru$_2$Re**. This result indicated that the Mizoroki–Heck reaction could proceed even in the absence of phosphine ligands. **Ru$_2$Re** was chromatographically isolated by SEC, giving an isolated yield of 25%. However, interestingly, the SEC chart measured after the reaction without phosphine ligands was quite different from that of the case with phosphine ligands and air

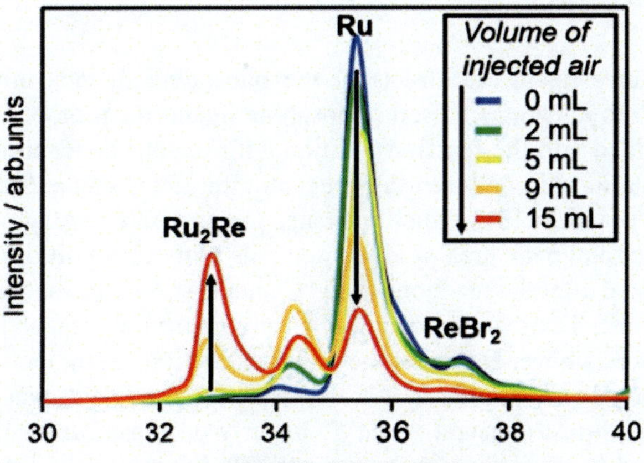

Fig. 3.1 SEC charts measured after heating the reaction solution at 65 °C under Ar for 2 days (blue line). The other lines show the chromatograms measured after air injection and further heating for 2 days. The MeCN solution (1 mL) contained **Ru** (8 μmol), **ReBr$_2$** (4 μmol), Pd(OAc)$_2$ (4 μmol), AcONa (21 μmol) and Sphos (8 μmol)

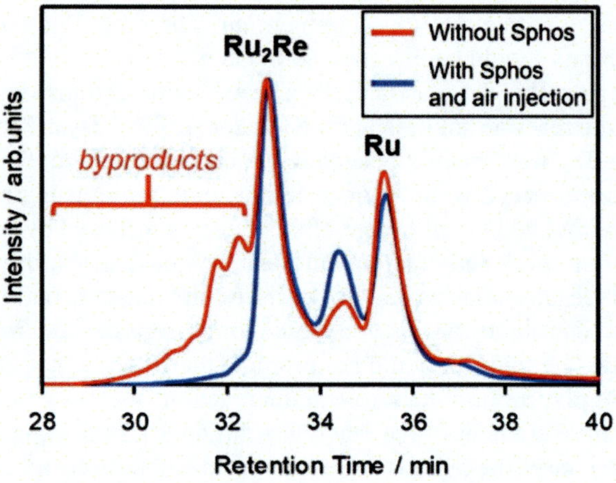

Fig. 3.2 SEC charts measured after the heating 3 days under Ar without phosphine ligands (red line) and with phosphine ligands followed by air injection (blue line)

injection (Fig. 3.2 blue line). In the case using no phosphine ligand, many peaks having shorter retention times than that of the peak attributed to $\mathbf{Ru_2Re}$ (retention time: 33 min) were observed. This result shows that byproducts having a larger molecular size than that of $\mathbf{Ru_2Re}$ were produced only when phosphine ligands were not used.

3.2.2 *Concentration Effect of Phosphine Ligands on the Reaction Abilities of the Mizoroki–Heck Reaction*

In order to investigate the effects of the phosphine ligands in the Pd catalyst, Sphos, which is known as a useful phosphine ligand for promoting coupling reactions, was added into the reaction solution with various concentrations. Figure 3.3 shows the relationship between the concentration ratios of the Sphos to $Pd(OAc)_2$ ([Sphos]/[Pd]) and S_1 [%], which indicates the area ratio of the peak attributed to $\mathbf{Ru_2Re}$ to the total peak area of all compounds with a Ru unit or units on the SEC charts measured after the reactions at 65 °C under Ar for 3 days (the detection wavelength: 299 nm). The reaction conditions were the same as those in the absence of Sphos described above. Note that the molar extinction coefficients of **Ru** and each Ru unit in $\mathbf{Ru_2Re}$ are similar at 299 nm. When [Sphos]/[Pd] was from 0 to 1, the S_1 value was almost constant while S_1 linearly decreased from [Sphos]/[Pd] was between 1 and 2. According to the reported studies using the Pd catalysts for the coupling reactions, one equivalent of added phosphine should be oxidized to reduce Pd(II) to Pd(0) [8]. Therefore, in the case of [Sphos]/[Pd] $\leq$ 1, there was probably no Sphos that could coordinate to the Pd(0) center in the reaction solution. It is

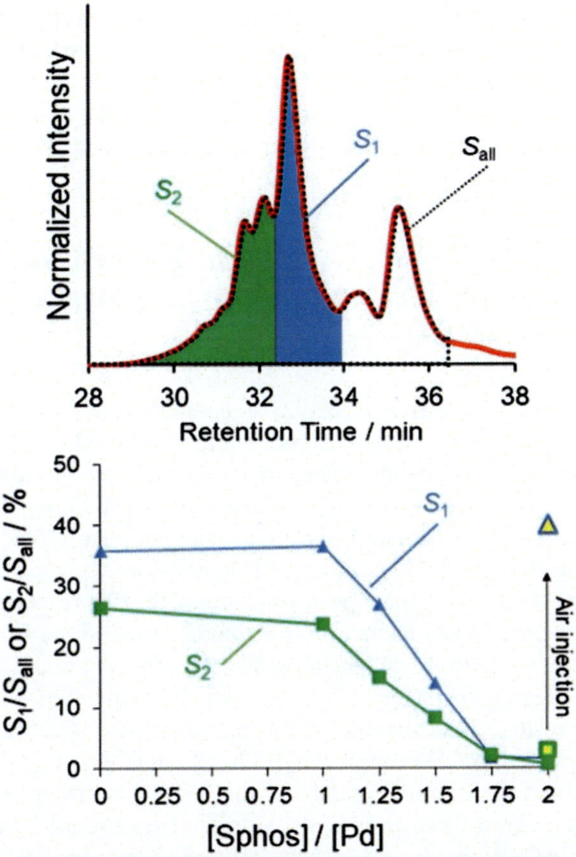

Fig. 3.3 Relationship between [Sphos]/[Pd] and peak area ratios of the peak attributed to $\mathbf{Ru_2Re}$ (blue triangles) or byproducts (green squares) to the sum of the peaks of all compounds having the Ru unit(s) on SEC charts measured after the reaction under Ar. The yellow triangle ($\mathbf{Ru_2Re}$) and square (byproducts) show the peak area ratios after air injection followed by further heating for 2 days

understandable that the results under these reaction conditions were similar to those in the absence of Sphos. On the other hand, the S_1 value became smaller in the case of [Sphos]/[Pd] > 1, where some Pd(0) should have a Sphos ligand or ligands. These results clearly indicate that the coordination of Sphos to the Pd(0) species suppressed the Mizoroki–Heck reaction, and the "*real catalyst*" for the Mizoroki–Heck reaction should be Pd(0) compounds without any phosphine ligands. Since S_1:S_2, the ratio of the product of the Mizoroki–Heck reaction (S_1) to byproducts (S_2), were quite similar regardless of the [Sphos]/[Pd] (Fig. 3.3), both the product and the byproducts should be catalyzed by the same Pd species.

As described above, some air was required to promote the Mizoroki–Heck reaction in the presence of phosphine ligands. Therefore, some air was injected into the reaction vessel by using a syringe after heating the reaction solution containing two equivalents of Sphos for 2 days under Ar. The reaction proceeded after injecting some air and **Ru_2Re** was detected as the main product on the SEC chart measured after the reaction (Fig. 3.3 yellow triangle). The S_1 reached around 40%, which was similar to that without Sphos, even though the increase of S_2 was not observed.

3.2.3 Investigation of the Byproducts Observed During the Mizoroki–Heck Reaction in the Absence of Phosphine Ligands

To investigate the side reaction, an excess amount of **Ru** (4 equivalents of **Ru** to one of **$ReBr_2$**) was added to the reaction solution (Fig. 3.4 green line). The peak attributed to **Ru_2Re** became larger, and the selectivity of the reaction was improved compared to the case using only two equivalents of **Ru** (Fig. 3.4 red line). The isolated yield became higher (31%) than that using 2 equivalents of **Ru** (25%), and the surplus of **Ru** was recollected by 71%. Since the side reaction was suppressed by using an excess amount of compounds having a vinyl group, the side reaction was supposed to occur between compounds having bromo groups. Some of the fractions containing the byproducts, which were collected by SEC (Fractions **1** and **2** in Fig. 3.5), were analyzed by electron spray ionization mass-spectroscopy (ESI–MS). According to the ESI–MS results, these byproducts were attributable to tetra- or heptanuclear complexes having more $[Re(bpy)(CO)_2(PPh_3)_2]^+$ moieties (bpy = 2,2′-bipyridine) than **Ru_2Re**, and the diimine ligands on the Re(I) moieties were supposed to be linked directly. From these results, the side reaction should be a homo-coupling reaction among C–Br bonds caused by reductive elimination after oxidative addition and transmetalation.

3.2.4 Confirmation of the Side Reaction: Selective Homo-coupling Reactions

To confirm the procedure of the homo-coupling reaction, a reaction using **ReBr(Et)** having one bromo group, $Pd(OAc)_2$, and AcONa was performed (Eq. 3.1). Although the turnover number of the Pd catalyst (< 1) is not high, the homo-coupling reaction underwent as $Pd(OAc)_2$ was charged. Since no side reaction was observed in the ESI-mass spectra, it is supposed that the homo-coupling reaction proceeded selectively (Fig. 3.6a). The purification was achieved by using ion-exchange chromatography, and the isolation yield reached 71%. From the ^1H NMR measurement of obtained **Re–Re(Et)**, only seven peaks attributed to seven kinds of protons on the diimine ligands were observed (Fig. 3.6b). From these results, only C–Br bonds worked as active spots of the homo-coupling. In the case using **ReBr(Ph)**, the dimer products were also obtained using the same procedure with a 28% isolated yield.

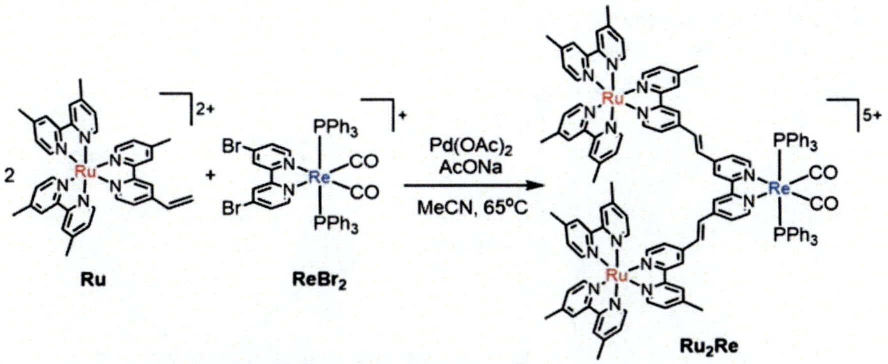

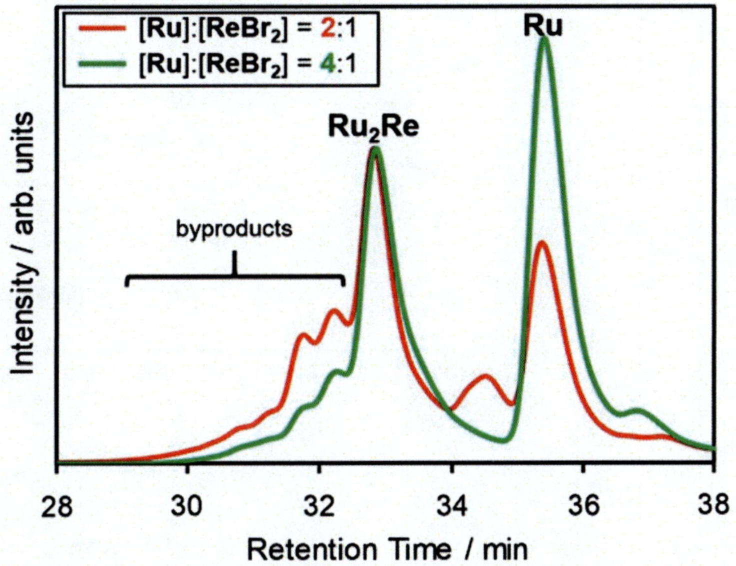

Fig. 3.4 SEC charts measured after the reactions with four equivalents (green line) or two equivalents of **Ru** (red line) to one of **ReBr**$_2$

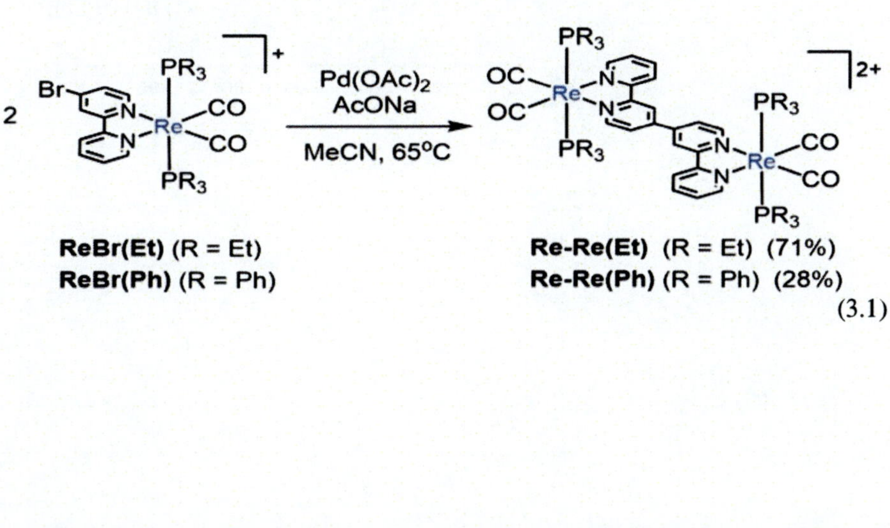

(3.1)

Fig. 3.5 ESI-mass spectra of byproducts from fractions **1** and **2** separated by using SEC

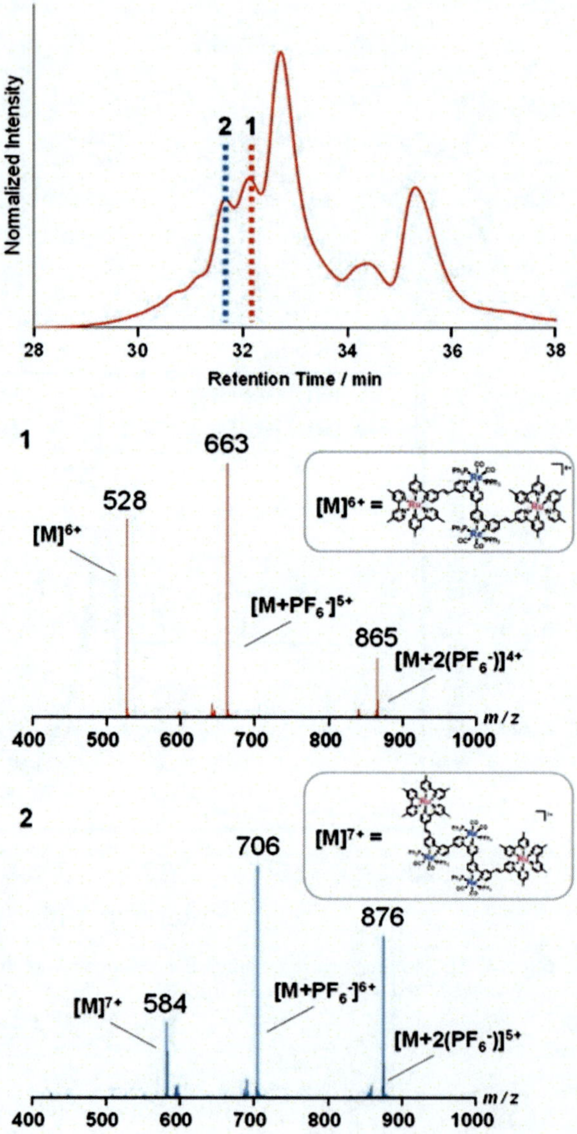

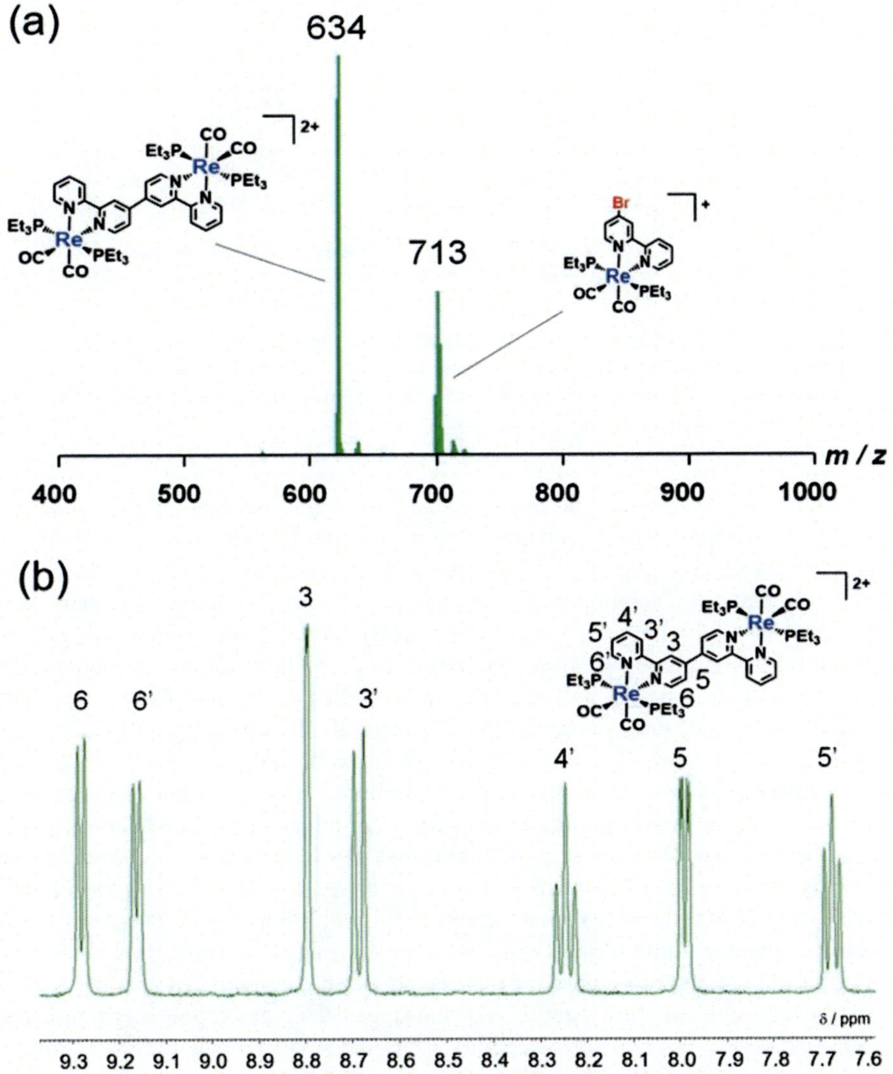

Fig. 3.6 **a** ESI-mass spectrum measured after the reaction for the synthesis of **Re–Re(Et)** and **b** ^{1}H NMR spectrum of the bridging ligand on **Re–Re(Et)**

3.2.5 Particle Size Measurements of Pd Compounds During the Mizoroki–Heck Reaction

Some literature reports that phosphine-free coupling reactions are often catalyzed by Pd nanoparticles, i.e., heterogeneous Pd catalysts [9]. Therefore, the particle size of the Pd catalyst in the reaction solution was measured by diffusion light

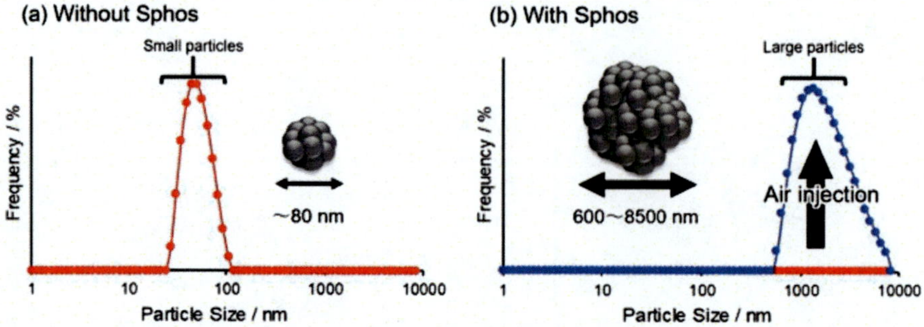

Fig. 3.7 The diffusion light scattering (DLS) results of the MeCN solutions containing **a** $Pd(OAc)_2$ and AcONa, or **b** $Pd(OAc)_2$, AcONa, and Sphos (red line) after heating overnight at 65 °C under Ar atmosphere. The blue line in (**b**) shows the DLS result measured after heating for 8 h with air injection into the solution

scattering (DLS). Firstly, $Pd(OAc)_2$ and AcONa were dissolved in MeCN bubbled by N_2. The solution was poured into a quartz cell after filtration through a membrane filter (Millex LG 0.20 μm) and bubbled by Ar. Before measuring particle size, the cell containing the solution was kept at 65 °C for 3 h. In the measurement, some nanoparticles, of which sizes are approximately 80 nm, were observed (Fig. 3.7a). Since no nanoparticle was observed before heating the solution, the nanoparticle was produced during the heating. The nanoparticles might be stabilized in a polar solvent containing many ions, e.g., AcO^-. Secondly, $Pd(OAc)_2$, AcONa, and Sphos were dissolved in MeCN bubbled by N_2. The solution was poured into a quartz cell after filtration through a membrane filter and bubbled by Ar. Before measuring the particle size, the cell containing the solution was kept at 65 °C for 8 h; no significant nanoparticle was observed (Fig. 3.7b red line). Some air was then injected into the cell by using an air pump for 8 h, and particle size was measured again. Particles with sizes of 600–8500 nm were observed (Fig. 3.7b blue line). These sizes were much larger than those formed with only $Pd(OAc)_2$ and AcONa as described above. From these results, it is suggested that the observed nanoparticles possibly work as catalysts for the coupling reaction, and the size of the nanoparticle might affect the catalytic ability and reaction selectivity. Only the small Pd nanoparticles might be able to catalyze both the Mizoroki–Heck reaction and the homo-coupling reaction.

3.2.6 Improvement of the Homo-coupling Reaction Using Metal Indium as an Electron Donor

The turnover number of the catalytic cycle for the homo-coupling reaction using $Pd(OAc)_2$, mentioned above, was less than one. This result might indicate that the homo-coupling reaction proceeded stoichiometrically. Scheme 3.2 shows the proposed reaction mechanism of the homo-coupling reaction under the reaction

Scheme 3.2 The proposed reaction mechanism of the homo-coupling reaction shown in Eq. 3.1

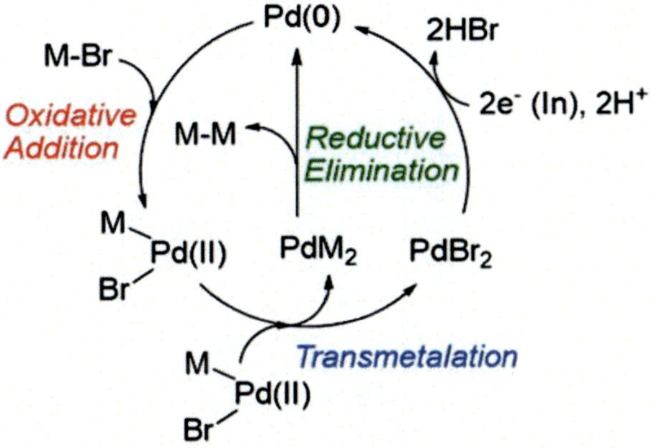

condition illustrated in Eq. 3.1 [4]. After oxidative addition of the complexes possessing a bromo group (**M-Br**) to the Pd(0) species and transmetalation between these adducts, two kinds of Pd(II) species that have two complexes (**PdM_2**) or two bromide ions (**$PdBr_2$**), should be produced. Pd(0) species should be recovered by reductive elimination of two complexes from **PdM_2** to produce targeted complexes **M-M**. On the other hand, **$PdBr_2$** might require electrons to regenerate Pd(0) species accompanying the release of two bromide ions. This might explain why the turnover number of the homo-coupling reaction became low. Therefore, according to the reaction mechanism, the homo-coupling reaction should be enhanced by adding some electron donors. It has been reported that metal indium (In) is a suitable electron donor to promote homo-coupling reactions. To conduct the homo-coupling reaction of the metal complexes catalytically, some In was added to the reaction solution. An MeCN solution (1 mL) containing **ReBr(Ph)** (51 mg, 44 μmol), In (23 mg, 200 μmol), AcONa (9.1 mg, 110 μmol) and ten mol% of $Pd(OAc)_2$ (1.0 mg, 4.4 μmol) was refluxed for 3 days. During the reaction, the reaction solution was stirred strongly to break the surface of the In-grain. The chart of SEC measured after the reaction is shown in Fig. 3.8 (red line). The peak attributed to the targeted dinuclear complex became larger than that for the reaction without In [Fig. 3.8 (blue line)]. The conversion yield was estimated at approximately 75% from the SEC chart, and the turnover number of the coupling reaction based on the amount of $Pd(OAc)_2$ (10 mol%) became larger than seven. This result clearly showed that electron donors for Pd(0) production, i.e., In, could promote the homo-coupling reaction. According to the ESI-mass spectrum measured after the reaction indicated that mononuclear complexes in the reaction solution, which could not be converted to dinuclear complexes, were all debrominated.

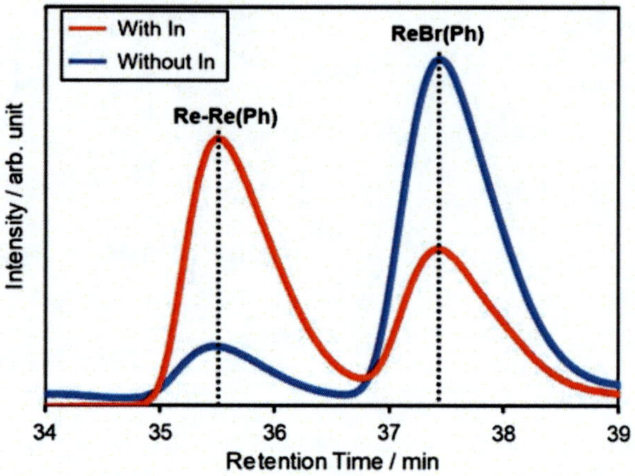

Fig. 3.8 The SEC charts measured after the homo-coupling reaction for **Re–Re(Ph)** in the presence of Indium (red line) or in the absence of Indium (blue line)

3.2.7 Strategies for Selective Mizoroki–Heck Reactions and Homo-coupling Reactions

As mentioned above, the Mizoroki–Heck reaction and the homo-coupling could proceed when using no phosphine ligands, and using an excess amount of **Ru** was useful to improve the reaction selectivity of the Mizoroki–Heck reaction. On the other hand, the Mizoroki–Heck reaction proceeded selectively when using some phosphine ligands and air injection processes after heating for a while. In the latter case, ^{31}P NMR measurement and ESI-mass analysis of the reaction solution after injecting some air revealed that all of the phosphine ligands were oxidized. Since the reaction proceeded continuously even after oxidizing the phosphine ligands, Pd compounds without phosphine ligands should work as catalysts for the Mizoroki–Heck reaction. Actually, the Mizoroki–Heck reaction proceeded without phosphine ligands and air injection. When the oxidized phosphine ligands were used instead of phosphine ligands, the homo-coupling reactions could not be suppressed, and SEC charts recorded after the reactions became almost the same as the red line in Fig. 3.2. Besides, even when using 2 equivalents of Sphos, the homo-coupling reactions proceeded under *air* or under Ar with introducing some air as soon as the heating was started. From these results, air should be injected after heating the reaction solution containing phosphine ligands for several hours to suppress the homo-coupling. It is supposed that not oxidized phosphine ligands but phosphine ligands should be required to suppress the homo-coupling reaction *only in the initial stage* of the reaction before the air injection, even though the phosphine ligands should be oxidized to remove them from Pd(0) compounds and to proceed the Mizoroki–Heck reaction. It is noteworthy that the phosphine oxide itself did not affect the reaction selectivity; these curious phenomena probably indicate that heterogeneous Pd compounds catalyzed the coupling reactions. Based on the mechanistic investigation

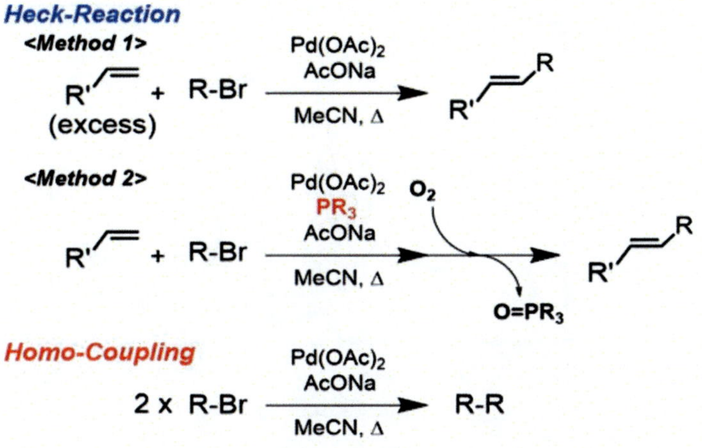

Scheme 3.3 Synthetic scheme for the selective Mizoroki–Heck reaction and the homo-coupling

described above, I can propose new and convenient methods for synthesizing various multinuclear complexes using the Pd compounds (Scheme 3.3).

There are two ways for improving the yield of the hetero-coupling products by the Mizoroki–Heck reaction,

Method 1: An excess amount of the complex with a vinyl group compared to the complexes with bromo groups, was used with only $Pd(OAc)_2$, and AcONa in the absence of the phosphine ligands.

Method 2: A phosphine ligand such as Sphos was added to the reaction solution, and then some air was injected into the reaction vessel after heating the solution for several hours. The solution was continuously heated for a few days.

The homo-coupling reaction between complexes with bromo groups proceeds selectively using only $Pd(OAc)_2$ and AcONa. This homo-coupling reaction could possibly be promoted in the presence of metal indium.

3.2.8 Synthesis of Novel Multinuclear Complexes Using a Combination of the Mizoroki–Heck Reaction and the Homo-coupling Reaction

By using some combinations of these methods, I successfully synthesized some new multinuclear complexes shown in Eqs. 3.2–3.4. Of note, these products, i.e., **RuRe-Br**, **RuRe-ReRu,** and **OsReRu**, cannot be obtained selectively using the ordinary methods, which require stepwise complexation between bridging ligands and metal complexes. In the synthesis of **RuRe-Br** (Eq. 3.2), NOT **Ru** but **$ReBr_2$** should be used with an excess amount to prevent the formation of the trinuclear complex (**Ru_2Re**) as a byproduct, which is produced by the further reactions between **Ru**

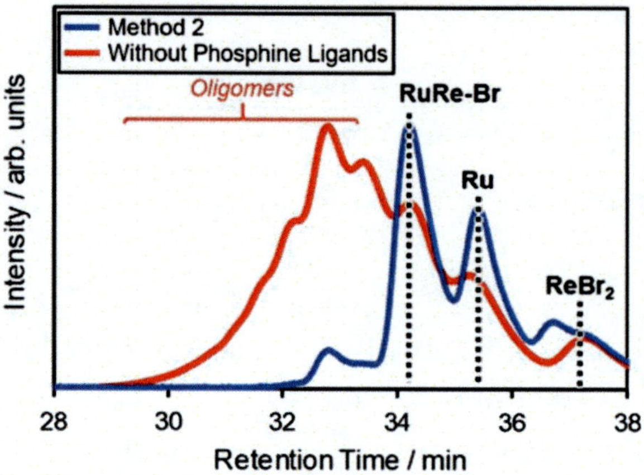

Fig. 3.9 The SEC charts measured after the reaction for the **RuRe-Br** using phosphine ligands and air injection (Method 2, blue line) or without phosphine ligands (red line)

and the targeted dinuclear complex (**RuRe-Br**). Therefore, I chose Method 2 for the synthetic method of **RuRe-Br** to suppress both polymerizations caused by homo-coupling and the production of **Ru_2Re**. An MeCN solution containing **Ru**, **$ReBr_2$**, $Pd(OAc)_2$, and AcONa was heated at 65 °C under an Ar atmosphere overnight. The air was gradually introduced by opening the three ways cock attached to the reaction vessel, and then the solution was heated for 1 week. The SEC chart of the reaction solution recorded after the reaction revealed that the targeted dinuclear complex was the main product (Fig. 3.9 blue line). From ESI–MS analysis, almost no debrominated products were detected. **RuRe-Br** was isolated by using ion-exchange chromatography and SEC, giving an isolated yield of 35% based on **Ru** used. On the other hand, in the absence of phosphine ligands, various multinuclear complexes were observed in the SEC charts measured after 3 days of heating (Fig. 3.9 red line).

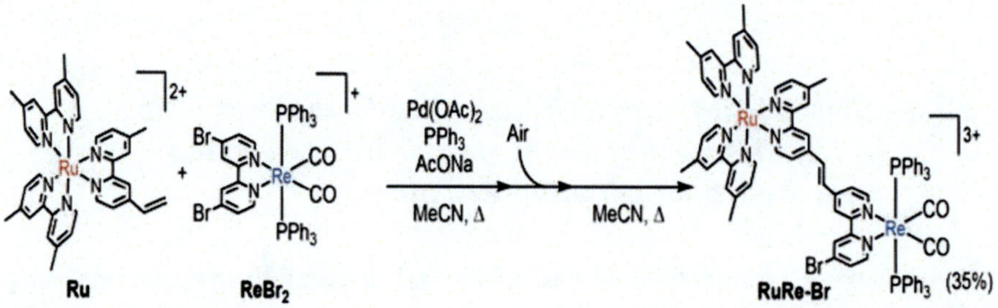

(3.2)

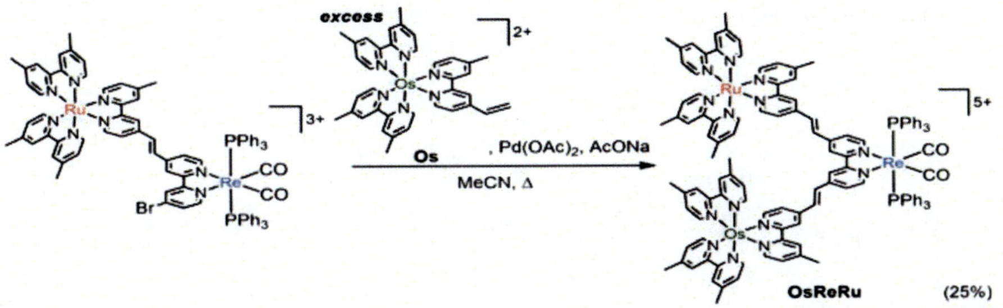

(3.3)

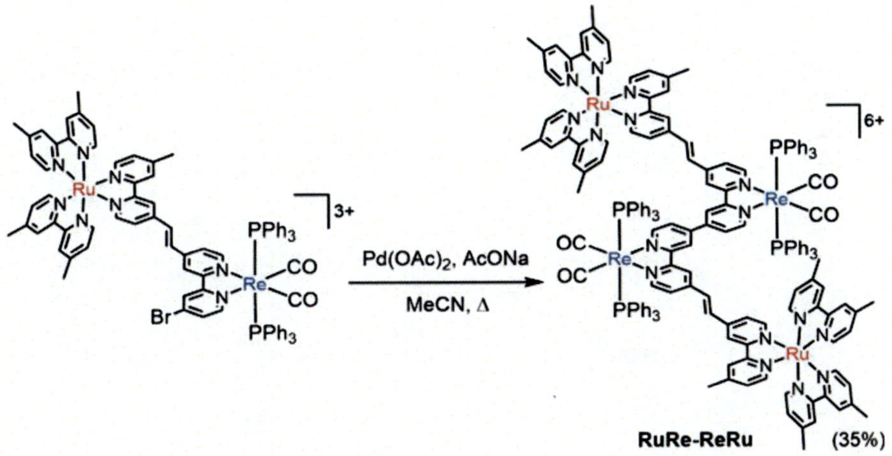

(3.4)

For synthesizing **RuRe-ReRu**, an MeCN solution containing **RuRe-Br**, $Pd(OAc)_2$, and AcONa was heated for 2 days followed by twice repeating the following processes: loading of additional $Pd(OAc)_2$ and heating for 2 days. The targeted tetranuclear complex was obtained as the main product with almost no byproducts, which was clearly demonstrated by SEC and ESI–MS measurements (Fig. 3.10a); only **RuRe-ReRu** and the starting complexes **RuRe-Br** were detected. The isolation was achieved using ion-exchange chromatography and SEC with an isolated yield of 35% based on **RuRe-Br** used. It should be noted that the homo-coupling reaction can be applied even for coupling between dinuclear complexes with high selectivity.

Another trinuclear complex with three different kinds of metal ions, i.e., **OsReRu**, was also successfully synthesized using **RuRe-Br** and an excess amount of **Os** (Method 1). Under a similar reaction condition for the syntheses of **RuRe-Br** and **RuRe-ReRu**, the Mizoroki–Heck reaction proceeded selectively between **Os** and **RuRe-Br**. In the SEC chart measured after the reaction (Fig. 3.10b), three major peaks, which are attributed to the targeted product **OsReRu** and two starting complexes, were observed. **OsReRu** was isolated by using ion-exchange chromatography and SEC with an isolated yield of 25%. It is noteworthy that **OsReRu** had

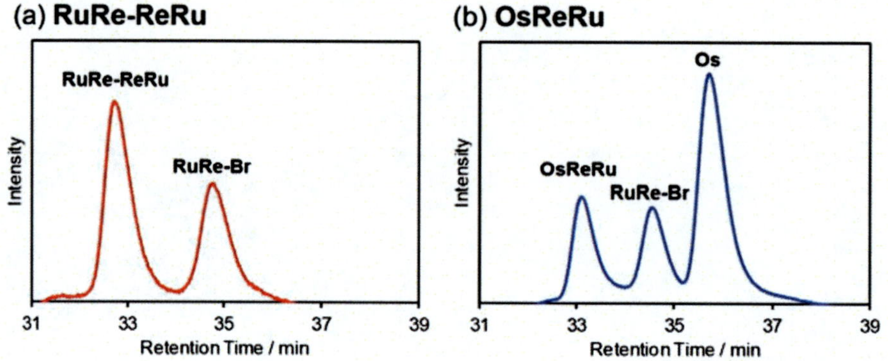

Fig. 3.10 The SEC charts measured after the reaction for the synthesis of **a RuRe-ReRu** and **b OsReRu**

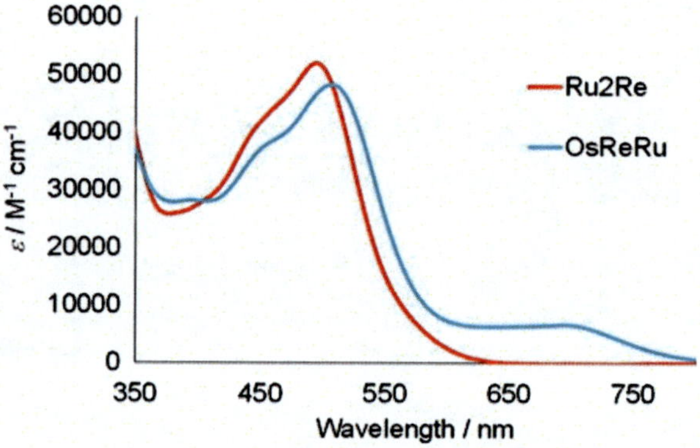

Fig. 3.11 UV–vis absorption spectra of **Ru$_2$Re** (red line) and **OsReRu** (blue line) measured in MeCN

strong absorption over a very wide range of UV–visible regions up to around 800 nm (Fig. 3.11).

3.2.9 Photophysical Properties of OsReRu

The photophysical properties of **OsReRu** obtained using stepwise Mizoroki–Heck reactions were investigated because intramolecular energy transfer might occur from the Ru(II) and Re(I) units to the Os(II) unit [10]. Tables 3.1 and 3.2 show photophysical properties of **OsReRu**, **Ru$_2$Re**, **Os$_2$Re**, and the corresponding mononuclear complexes (Fig. 3.12). The ^{1}MLCT absorption bands of **Os$_2$Re**, **OsReRu**, and **Ru$_2$Re** were observed at λ_{max} = 520, 506, and 494 nm, respectively, and these were red-shifted by more than 30 nm compared to the ^{1}MLCT absorption bands

Table 3.1 Absorption properties of $\mathbf{Os_2Re}$, **OsReRu**, $\mathbf{Ru_2Re}$, and corresponding mononuclear complexes measured in MeCN

Complex	λ_{abs}/nm ($\varepsilon/10^3\ M^{-1}\ cm^{-1}$)		
	π-π*	^{1}MLCT	^{3}MLCT
$\mathbf{Os_2Re}$	290 (131)	520 (47.2)	693 (11.5)
OsReRu	289 (149)	506 (48.2)	694 (6.64)
$\mathbf{Ru_2Re}$	288 (150)	494 (52.1)	
Os	291 (108)	464 (19.7)	600 (5.51)
Ru	288 (73.1)	462 (16.3)	
Re	309 (22.2)	430 (4.29)	

Table 3.2 Emission properties of $\mathbf{Os_2Re}$, **OsReRu**, $\mathbf{Ru_2Re}$ and corresponding mononuclear complexes[a]

Complex	Φ_{em}	τ_1/ns (A_1/%)	τ_2/ns (A_2/%)	τ_3/ns (A_3/%)
$\mathbf{Os_2Re}$[b]	0.005	854 (56)[e]	22 (44)[e]	
OsReRu[b]	0.006	914 (41)[e]	17 (52)[e]	93 (7)[e]
$\mathbf{Ru_2Re}$[c]	0.02	1310 (82)[f]	610 (18)[f]	
Os[b]	0.01	31 (100)[e]		
Ru[c]	0.10	1370 (100)[f]		
Re[d]	0.08	1561 (100)[f]		

[a] Measured in MeCN at 25 °C. [b] Excitation wavelength: 510 nm. [c] Excitation wavelength: 456 nm. [d] Excitation wavelength: 444 nm. [e] Detected at 750 nm. [f] Detected at each wavelength of the emission maximum

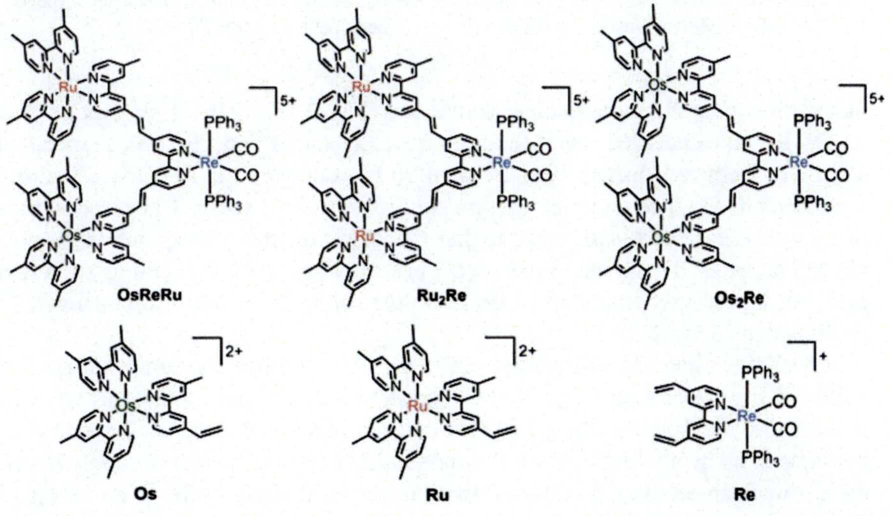

Fig. 3.12 Structure of $\mathbf{Os_2Re}$, **OsReRu**, $\mathbf{Ru_2Re}$, and the corresponding mononuclear complexes

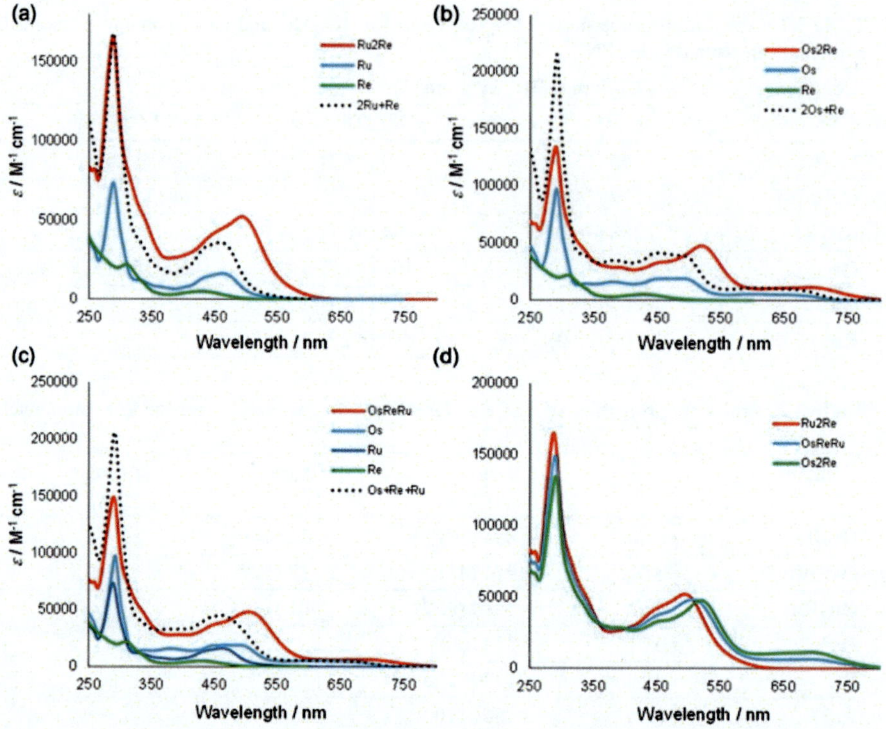

Fig. 3.13 **a** Absorption spectra of **Ru_2Re** (red), **Ru** (blue), **Re** (green), and the 2:1 summation of **Ru** and **Re** (dotted line). **b** Absorption spectra of **OsReRu** (red), **Os** (blue), **Ru** (dark blue), **Re** (green), and the summation of **Os**, **Ru**, and **Re** (dotted line). **c** Absorption spectra of **Os_2Re** (red), **Os** (blue), **Re** (green), and the 2:1 summation of **Os** and **Re** (dotted line). **d** Absorption spectra of **Ru_2Re** (red), **OsReRu** (blue), and **Os_2Re** (green). The solvent was MeCN

of the corresponding mononuclear complexes (Fig. 3.13). ^{3}MLCT absorption bands of **Os_2Re** and **OsReRu** were observed at $\lambda_{max} = 693$ and 694 nm, respectively, and these were red-shifted by more than 90 nm compared to the ^{3}MLCT absorption bands of **Os**. The molar extinction coefficients of the ^{1}MLCT absorption bands of the trinuclear complexes were higher than those of the corresponding mononuclear complexes. These phenomena were probably because the π conjugation in the bridging ligand was extended to decrease the π^* level of the bridging ligand, as described in Chap. 2.

All of the trinuclear complexes emitted light, although the emission quantum yields of trinuclear complexes were much lower than those of the corresponding mononuclear complexes (Table 3.2). The emission decays of the trinuclear complexes could be fitted using double or triple exponential functions. This is probably because each trinuclear complex had two or three different kinds of units. Even though the emission quantum yields of the trinuclear complexes were much lower, their emission lifetimes were similar to those of the corresponding mononuclear complexes.

Therefore, only the radiative decay constants of the trinuclear complexes should be slower than those of the mononuclear complexes. The slower radiative decay constants might indicate that the π-π^* characteristics of the lowest ^{3}MLCT excited states became stronger in the trinuclear complexes than those in the mononuclear complexes. This is supported by the fact that the molar extinction coefficients of the ^{1}MLCT absorption of the trinuclear complexes were larger than those of the mononuclear complexes, as described in Chap. 2.

The τ_1 and τ_2 of **Os$_2$Re** could be respectively attributed to emission lifetimes of the Re unit and the Os units because of the similarities with those of the corresponding mononuclear complexes. The relatively large τ_1 value suggested that the energy transfer from the Re(I) unit to the Os(II) units should be a relatively slow process. The τ_1 and τ_2 of **OsReRu** were similar to those of **Os$_2$Re**, while the τ_3 was completely different from the lifetimes of **Os$_2$Re** and **Ru$_2$Re**. These results might indicate that the energy transfer from the Ru unit to the Os unit might take place more rapidly.

3.3 Conclusions

In this chapter, I clarified the role of air in the coupling reactions, which could promote the Mizoroki–Heck reaction, by investigating the concentration effect of phosphine ligands on the reaction. The Mizoroki–Heck reaction could be catalyzed by Pd compounds possessing no phosphine ligands; therefore, some air was required to oxidize and remove the phosphine ligands from Pd compounds. The existence of the phosphine ligands in the initial stage of the reaction might control the particle size of Pd particles to catalyze the Mizoroki–Heck reaction selectively. On the other hand, in the absence of phosphine ligands, not only the Mizoroki–Heck reaction but also the homo-coupling reaction proceeded simultaneously. In this case, the selectivity for the Mizoroki–Heck reaction was improved by using excess amounts of the complexes possessing vinyl groups. The homo-coupling reaction selectively proceeded under the same reaction conditions in the absence of complexes with vinyl groups. Using a combination of these reactions, novel multinuclear complexes, which are difficult to be synthesized by the ordinary methods, were successfully obtained. In particular, the obtained trinuclear complexes with an Os(II) unit showed strong absorption abilities over a wide range of visible regions.

3.4 Experimental

General experimental procedures have been described in detail in the following journal articles published previously:

Y. Yamazaki, O. Ishitani, "Selectivity control between Mizoroki–Heck and homo-coupling reactions for synthesizing multinuclear metal complexes: unique addition effects of tertiary phosphines and O_2", *Dalton Trans.*, **2017**, *46*, 4816–4823.

3.4.1 Materials

All reagents were of reagent-grade quality and were used without further purification. [Re(Brbpy)$(CO)_2(PEt_3)_2$](PF_6) (**ReBr(Et)**), [Re(Brbpy)$(CO)_2(PPh_3)_2$](PF_6) (**ReBr(Ph)**), [Ru$(dmb)_2$(vbpy)]$(PF_6)_2$ (**Ru**) were prepared according to the reported methods described in Chap. 2 with some modifications.

3.4.2 Synthesis

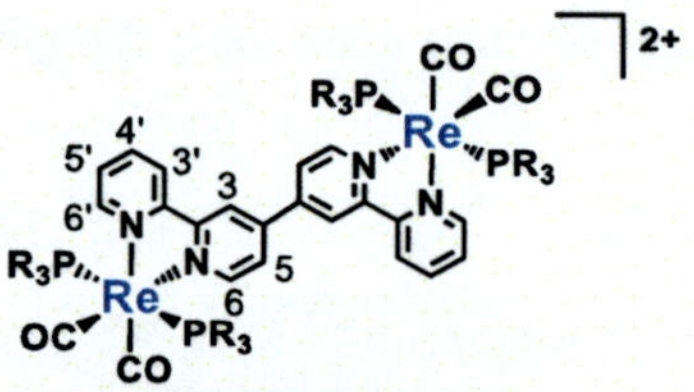

Re(Et)-Re(Et) (R = Et)
Re(Ph)-Re(Ph)(R= Ph)

Re(Et)-Re(Et): **ReBr(Et)** (10.0 mg, 11.6 μmol), $Pd(OAc)_2$ (1.8 mg, 8.0 μmol) and AcONa (3.4 mg, 41 μmol) were dissolved in MeCN (2 mL, bubbled by N_2). The solution was heated at 65 °C under Ar for 3 days. Additional $Pd(OAc)_2$ (1.8 mg, 8.0 μmol) was added to the solution and the solution was heated for another 3 days. After adding $Pd(OAc)_2$ (1.8 mg, 8.0 μmol) the solution was heated for 2 days. The solution was cooled to room temperature and was purified by ion-exchange chromatography [SP Sephadex C-25, eluent: MeCN-water (1:1 v/v) containing NH_4PF_6 (0–4 mM)]. Some portion of MeCN was evaporated under reduced pressure. The red precipitate was filtered off and washed with water and ether. Yield: 6.5 mg (72%). ^{1}H NMR (400 MHz, acetone-d_6): δ/ppm, 9.39 (d, $J = 5.6$ Hz, 2H, bpy-6), 9.27 (d, $J = 6.9$ Hz, 2H, bpy-6′), 8.91 (d, $J = 2.0$ Hz, 2H, bpy-3), 8.79 (d, $J = 8.2$ Hz, 2H, bpy-3′), 8.36 (dd, $J = 8.2$, 8.2 Hz, 2H, bpy-4′), 8.10 (dd, $J = 2.0$, 5.6 Hz, 2H, bpy-5), 7.79 (dd, $J = 6.9$, 8.2 Hz, 2H, bpy-5′), 1.63–1.57 (m, 24H, P–$\underline{CH_2}CH_3$), 0.98–0.91 (m, 36H, P–$CH_2\underline{CH_3}$). ^{31}P NMR (162 MHz, acetone-d_6): δ/ppm, -2.38 (s, 4P, -$\underline{P}Et_3$), – 144.0 (septet, 2P, $\underline{P}F_6^-$). FT-IR (in CH_2Cl_2) ν_{CO}/cm^{-1}: 1931, 1862. ESI–MS (in MeCN) *m/z*: 633 ([M-2PF_6^-]$^{2+}$). HRMS (ESI-TOF) *m/z*: [M-2PF_6^-]$^{3+}$ Calcd for $C_{48}H_{74}N_4O_4P_4Re_2$ 633.1875; Found 633.1891.

Re(Ph)-Re(Ph): **ReBr(Ph)** (13.0 mg, 11.3 μmol), $Pd(OAc)_2$ (1.0 mg, 4.5 μmol) and AcONa (4.6 mg, 57 μmol) were dissolved in MeCN (1 mL, bubbled by N_2). The solution was heated at 65 °C under Ar for 2 days. Additional $Pd(OAc)_2$ (1.0 mg, 4.5 μmol) was added to the solution and the solution was heated for another 2 days. After adding $Pd(OAc)_2$ (1.0 mg, 4.5 μmol) the solution was heated for 2 days. The solution was cooled to room temperature and was purified by ion-exchange chromatography (SP Sephadex C-25, eluent: MeCN-water (1:1 v/v) containing NH_4PF_6 (0–4 mM)). Some portion of MeCN was evaporated under reduced pressure. The red precipitate was filtered off and washed with water and ether. Yield: 3.4 mg (28%). ^{1}H NMR (400 MHz, acetone-d_6): δ/ppm, 8.29 (d, J = 4.6 Hz, 2H, bpy-6), 8.27 (d, J = 7.8 Hz, 2H, bpy-6′), 8.11 (d, J = 1.6 Hz, 2H, bpy-3), 7.88 (dd, J = 6.3, 7.0 Hz, 2H, bpy-4′), 7.76 (d, J = 6.3 Hz, 2H, bpy-3′), 7.39–7.28 (m, 30H, 6PPh$_3$), 6.96 (dd, J = 7.0, 7.8 Hz, 2H, bpy-5′), 6.73 (dd, J = 1.6, 4.6 Hz, 2H, bpy-5). ^{31}P NMR (162 MHz, acetone-d_6): δ/ppm, 22.3 (s, 4P, -PPh$_3$), − 143.6 (septet, 2P, PF_6^-). FT-IR (in CH_2Cl_2) ν_{CO}/cm^{-1}: 1939, 1871. ESI–MS (in MeCN) *m/z*: 922 ([M-2PF$_6^-$]$^{2+}$). HRMS (ESI-TOF) *m/z*: [M-2PF$_6^-$]$^{3+}$ Calcd for $C_{96}H_{74}N_4O_4P_4Re_2$ 922.1896; Found 922.1922.

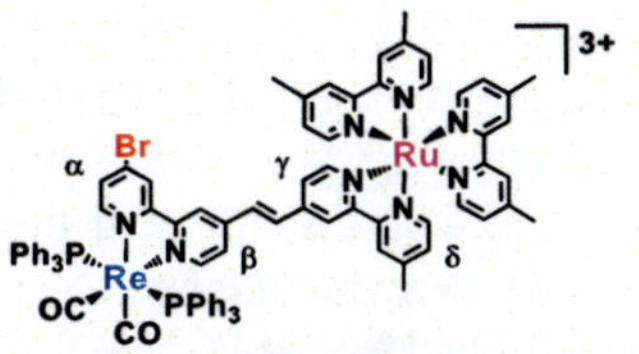

RuRe-Br

RuRe-Br: **ReBr$_2$** (30 mg, 0.024 mmol), **Ru** (12 mg, 0.012 mmol), Pd(OAc)$_2$ (2.7 mg, 0.012 mmol), PPh$_3$ (9.9 mg, 0.024 mmol) and AcONa (5.1 mg, 0.062 mmol) were dissolved in MeCN (3 mL, degassed by N_2). The solution was heated at around 75 °C under Ar for 10 h (in dim light). The atmosphere was gradually changed to air by opening the three ways cock, and the solution was heated for 1 week. The solvent was removed in vacuo, and a red solid was obtained. The solid was purified by ion-exchange chromatography (CM Sephadex C-25, eluent: MeCN-water (1:1 v/v) containing NH_4PF_6 (0–16 mM)) and size exclusion chromatography. The solvent of the obtained red solution was evaporated. The residue was dissolved in CH_2Cl_2 and washed with water containing NH_4PF_6 twice. After evaporating, the obtained red solid was dissolved in MeOH, and a concentrated methanol solution of NH_4PF_6 was added. Some water was added to the solution, and a portion of MeOH was evaporated under reduced pressure. After filtration, the red solid was collected and washed with water and Et$_2$O and dried in a vacuum at 60 °C. Yield: 9.1 mg (35%).^{1}H NMR (400 MHz, acetone-d_6): δ/ppm, 8.97 (s, 1H, δpy-3), 8.73 (s, 1H, γpy-3), 8.69 (s, 4H, dmb-3), 8.60 (s, 1H, αpy-3), 8.51 (s, 1H, βpy-3), 8.07 (d, J = 5.4 Hz, 1H, δpy-6), 8.04 (d, J = 6.0 Hz, 1H, γpy-6), 8.60 (d, J = 6.0 Hz, 1H, αpy-6), 8.51 (d, J = 5.6 Hz, 1H, βpy-6), 7.80 (m, 2H, –CH=CH–), 7.87–7.83 (m, 4H, dmb-6), 7.64 (d, J = 5.6 Hz, 1H, βpy-5), 7.43–7.31 (m, 35H, 6Ph, dmb-5, γpy-5) 7.18 (d, J = 6.0 Hz, αpy-5), 7.17 (d, J = 5.4 Hz, δpy-5), 2.61 (s, 3H, δpy-CH$_3$), 2.58 (s, 12H, dmb-CH$_3$). ^{31}P NMR (162 MHz, acetone-d_6): δ/ppm, 21.8 (s, 2P, -PPh$_3$), − 143.5 (septet, 3P, PF_6^-). FT-IR (in CH_2Cl_2) ν_{CO}/cm^{-1}: 1939, 1870. ESI–MS (in MeCN) *m/z*: 555 ([M-3PF$_6^-$]$^{3+}$). HRMS (ESI-TOF) *m/z*: [M-3PF$_6^-$]$^{3+}$ Calcd for $C_{85}H_{70}BrN_8O_2P_2ReRu$ 555.0989; Found 555.0999.

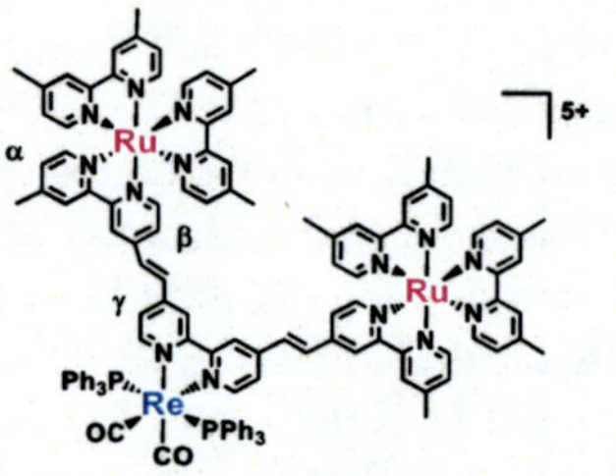

Ru₂Re

Ru_2Re:

Method (a)

$ReBr_2$ (30 mg, 0.024 mmol), **Ru** (47 mg, 0.049 mmol), $Pd(OAc)_2$ (5.5 mg, 0.024 mmol), PPh_3 (13 mg, 0.049 mmol) and AcONa (12 mg, 0.146 mmol) were dissolved in MeCN (4 mL, degassed by N_2). The solution was refluxed at around 75 °C under N_2 for 1 day (in dim light). A MeCN solution (4 mL, degassed by N_2) containing $Pd(OAc)_2$ (5.5 mg, 0.024 mmol) and PPh_3 (13 mg, 0.049 mmol) was added as a second charge *while opening the cap* and refluxed for 1 day. The solvent was removed in vacuo, and a black-red solid was collected. The solid was purified by size exclusion chromatography, and the solvent of the obtained red solution was evaporated. The residue was dissolved in CH_2Cl_2 and washed with water containing NH_4PF_6 twice. After evaporating, the obtained red solid was dissolved in MeOH, and a concentrated methanol solution of $NH_4PF_6$was added. Some water was added to the solution, and a portion of MeOH was evaporated under reduced pressure. After filtration, the red solid was collected and washed with water and Et_2O and dried in a vacuum at 60 °C. Yield: 23 mg (32%).

Method (b)

$ReBr_2$ (5.0 mg, 4.1 μmol), **Ru** (15.5 mg, 16.3 μmol), $Pd(OAc)_2$ (0.91 mg, 4.1 mmol) and AcONa (1.7 mg, 21 μmol) were dissolved in MeCN (1 mL, degassed by N_2). The solution was refluxed at around 65 °C under Ar for 3 days (in dim light). The solvent was removed in vacuo, and a black-red solid was collected. The solid was purified by size exclusion chromatography, and the solvent of the obtained red solution was evaporated. The residue was dissolved in CH_2Cl_2 and washed with water containing NH_4PF_6 twice. After evaporating, the obtained red solid was dissolved in MeOH, and a concentrated methanol solution of NH_4PF_6 was added. Some water was added to the solution, and a portion of MeOH was evaporated under reduced pressure. After filtration, the red solid was collected and washed with water and Et_2O and dried in a vacuum at 60 °C. Yield: 3.8 mg (31%).

^{1}H NMR (300 MHz, acetone-d_6): δ/ppm, 8.96 (s, 2H, βpy-3), 8.74 (s, 2H,αpy-3), 8.69 (s, 8H, dmb-3,3′), 8.51 (s, 2H, γpy-3), 8.10 (d, $J = 6.0$ Hz, 2H, γpy-6), 8.08 (d, $J = 6.5$ Hz, 2H, βpy-6), 7.98 (d, $J = 6.0$ Hz, 2H, γpy-5), 7.85–7.81 (m, 10H, αpy-6, dmb-6,6′), 7.72 (d, $J = 6.5$ Hz, 2H, βpy-5), 7.45–7.22 (m, 40H, – 6PPh_3, dmb-5,5′,–C<u>H</u>=C<u>H</u>–), 7.23 (d, J = 6.2 Hz, 2H, αpy-5), 2.65 (s, 6H, αpy-CH_3), 2.56 (s, 24H, dmb-CH_3). ^{31}P NMR (162 MHz, acetone-d_6): δ/ppm, 21.7 (s, 2P, -$\underline{P}Ph_3$), − 143.5 (septet, 5P, $\underline{P}F_6^-$). FT-IR (in CH_2Cl_2) ν_{CO}/cm^{-1}: 1939, 1870. ESI–MS (in MeCN) *m/z*: 450 ([M-5PF_6^-]$^{5+}$), 599 ([M-4PF_6^-]$^{4+}$), 847 ([M-3PF_6^-]$^{3+}$). HRMS (ESI-TOF) *m/z*: [M-5PF_6^-]$^{5+}$ Calcd for $C_{122}H_{106}N_{14}O_2P_2ReRu_2$ 450.3158; Found 450.3158, [M-4PF_6^-]$^{4+}$ Calcd for $C_{122}H_{106}F_6N_{14}O_2P_3ReRu_2$ 599.1357; Found 599.1362, [M-3PF_6^-]$^{3+}$ Calcd for $C_{122}H_{106}F_{12}N_{14}O_2P_4ReRu_2$ 847.1691; Found 847.1655.

Os(dmb)$_2$Cl$_2$: $(NH_4)_2[OsCl_6]$ (300.0 mg, 0.683 mmol) and 4,4′-dimethyl-2–2′-bipyridine (252.0 mg, 1.37 mmol) were dissolved in DMF (*ca.* 15 mL) and refluxed under Ar for 4 h. After cooling to room temperature, $Na_2S_2O_4$ (3.4 g) aqueous solution (15 mL) was added, and the solution was kept in a refrigerator for 12 h. The dark precipitate was collected by vacuum filtration and washed with water and Et_2O. The obtained solid was dried in a vacuum (65 °C). This solid was all used for the next reaction without further purification. Yield: 363.9 mg (85%).

[Os(II)(dmb)$_2$(vbpy)](PF_6)$_2$ (**Os**): Os(dmb)$_2$Cl$_2$ (101.6 mg, 0.1613 mmol) and 4-methyl-4′-vinyl-2–2′-bipyridine (vbpy, 31.7 mg, 0.162 mmol) was dissolved in EtOH-water mixed solution (28 mL, 5:2 v/v) and was refluxed under Ar for 5 h. After cooling to room temperature, an NH_4PF_6-saturated solution (ca. 15 mL) was added and the solution was kept in a refrigerator overnight. The dark precipitate was collected by vacuum filtration and washed with water and Et_2O. The obtained dark solid was purified by column chromatography on basic alumina [2nd fraction, eluent: CH_2Cl_2/MeOH (= 100:0–100:5)] and ion-exchange column chromatography (absorbent, SP Sephadex C-25; eluent, MeCN-water (1:1 v/v) containing NH_4PF_6, 0–8 mM). After removing portion of MeCN, a green-black precipitate was collected by vacuum filtration. The obtained green-black solid was washed with water and Et_2O, and dried in a vacuum (at 60 °C). Yield: 38.8 mg (24%). ESI–MS (in MeCN) *m/z*: 378 ([M-2PF_6^-]$^{2+}$), 900 ([M-PF_6^-]$^+$). ^{1}H NMR (400 MHz, acetone-d_6): δ/ppm, 8.80 (s, 1H, vbpy-3′), 8.71 (s, 1H, vbpy-3), 8.64 (s, 4H, dmb-3), 7.85 (d, J = 6.0 Hz, 1H, vbpy-6′), 7.80 (d, J = 6.0 Hz, 1H, vbpy-6), 7.79–7.73 (m, 4H, dmb-6), 7.46 (d, J = 6.0 Hz, 1H, vbpy-5′), 7.32–7.30 (m, 5H, dmb-5, vbpy-5), 6.92 (dd, J = 10, 18 Hz, 1H, bpy-C$\underline{H}$=CH$_2$), 6.32 (d, J = 18 Hz, 1H, bpy-CH=C$\underline{H}_2$(*trans*)), 5.62 (d, J = 10 Hz, 1H, bpy-CH=CH$_2$(*cis*)), 2.64 (s, 12 H, bpy-CH$_3$), 2.63 (s, 3H, vbpy-CH$_3$). HRMS (ESI-TOF) *m/z*: [M-2PF_6^-]$^{2+}$ Calcd for $C_{37}H_{36}N_6Os$ 378.1309; Found 378.1304.

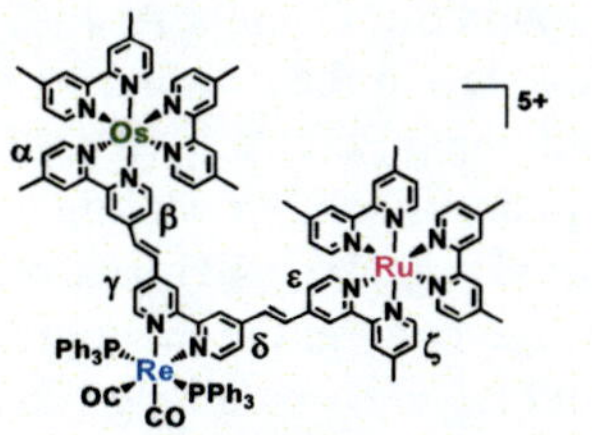

OsReRu

OsReRu: **RuRe-Br** (10 mg, 4.8 μmol), **Os** (10 mg, 9.5 μmol), Pd(OAc)$_2$ (0.53 mg, 2.4 μmol) and AcONa (0.98 mg, 12 μmol) were dissolved in MeCN (1 mL, degassed by N_2). The solution was refluxed at around 65 °C under Ar for 1 d (in dim light). $Pd(OAc)_2$ (0.53 mg, 2.4 μmol) was added to the solution as second charge. After heating 1 d, the solvent was removed under decreased pressure, and brown solid was collected. The solid was purified by ion-exchange column chromatography (absorbent, CM Sephadex C-25; eluent, MeCN-water (1:1 v/v) containing NH_4PF_6, 0–8 mM) and size exclusion chromatography. The solvent of the obtained brown solution was evaporated. The residue was dissolved in CH_2Cl_2 and washed with water containing NH_4PF_6 twice. After evaporating, the obtained red solid was dissolved in MeOH and a concentrated methanol solution of NH_4PF_6 was added. Some water was added to the solution and portion of MeOH was evaporated under reduced pressure. After filtration, red solid was collected and washed with water and Et_2O, and dried in vacuum at 60 °C. Yield: 3.6 mg (25%). ^{1}H NMR (400 MHz, acetone-d_6): δ/ppm, 8.96 (s, 1H, εpy-3), 8.91 (s, 1H, βpy-3), 8.74 (s, 1H, ζpy-3), 8.71 (s, 1H, αpy-3), 8.68 (s, 4H, dmb(Ru)-3,3′), 8.65 (s, 4H, dmb(Os)-3,3′), 8.51 (s, 1H, δpy-3), 8.50 (s, 1H, γpy-3), 8.09 (d, $J = 6.0$ Hz, 1H, δpy-6), 8.07–8.02 (m, 3H), 7.97 (d, $J = 6.0$ Hz, 2H, δpy-5), 7.87–7.62 (m, 16H), 7.56 (d, $J = 6.0$ Hz, 1H), 7.43–7.20 (m, 40H), 2.73 (s, 3H, αpy-CH_3), 2.65 (s, 3H, ζpy-CH_3), 2.64 (s, 12H, dmb(Os)–CH_3), 2.56 (s, 12H, dmb(Ru)–CH_3). ^{31}P NMR (162 MHz, acetone-d_6): δ/ppm, 21.8 (s, 2P, -$\underline{P}Ph_3$), − 143.5 (septet, 5P, $\underline{P}F_6^-$). FT-IR (in CH_2Cl_2) ν_{CO}/cm^{-1}: 1938, 1869. ESI–MS (in MeCN) *m/z*: 468 ([M-5PF_6^-]$^{5+}$), 621 ([M-4PF_6^-]$^{4+}$), 877 ([M-3PF_6^-]$^{3+}$). HRMS (ESI-TOF) *m/z*: [M-5PF_6^-]$^{5+}$ Calcd for $C_{122}H_{106}N_{14}O_2OsP_2ReRu$ 467.9265; Found 467.9257, [M-4PF_6^-]$^{4+}$ Calcd for $C_{122}H_{106}F_6N_{14}O_2OsP_3ReRu$ 621.1492; Found 621.1459.

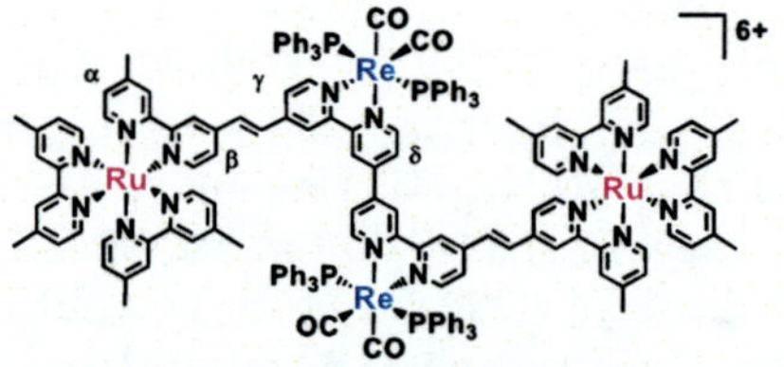

RuRe-ReRu

RuRe-ReRu: **RuRe-Br** (5.0 mg, 2.4 μmol), $Pd(OAc)_2$ (0.53 mg, 2.4 μmol) and AcONa (1.0 mg, 12 μmol) were dissolved in MeCN (0.5 mL, degassed by N_2). The solution was refluxed at around 65 °C under Ar for 2 days (in dim light). $Pd(OAc)_2$ (0.53 mg, 2.4 μmol) was added to the solution as second charge and the solution was heated for 2 days. Additional $Pd(OAc)_2$ (0.53 mg, 2.4 μmol) was added to the solution followed by 2 days heating. After cooling to room temperature, the solvent was removed under decreased pressure, and black-red solid was collected. The solid was purified by ion-exchange column chromatography (absorbent, CM Sephadex C-25; eluent, MeCN-water (1:1 v/v) containing NH_4PF_6, 0–8 mM) and size exclusion chromatography. The solvent of the obtained red solution was evaporated. The residue was dissolved in CH_2Cl_2 and washed with water containing NH_4PF_6 twice. After evaporating, the obtained red solid was dissolved in MeOH and a concentrated methanol solution of NH_4PF_6 was added. Some water was added to the solution and portion of MeOH was evaporated under reduced pressure. After filtration, red solid was collected and washed with water and Et_2O, and dried in a vacuum at 60 °C. Yield: 1.7 mg (35%). ^{1}H NMR (400 MHz, acetone-d_6): δ/ppm, 8.95 (s, 2H, βpy-3), 8.69 (s, 2H, αpy-3), 8.66 (s, 4H), 8.64 (s, 4H), 8.57 (d, J = 4.4 Hz, 2H, βpy-6), 8.43 (d, J = 5.2 Hz, 2H, αpy-6), 8.22 (d, J = 5.6 Hz, 1H), 7.92 (d, J = 6.0 Hz, 2H), 7.89 (s, 2H), 7.83 (d, J = 5.6 Hz, 4H), 7.82–7.78 (m, 6H), 7.87–7.83 (m, 4H, dmb-6), 7.64 (d, J = 4.4 Hz, 1H, βpy-5), 7.74–7.72 (m, 4H), 7.66–7.60 (m, 4H), 7.41–7.31 (m, 34H), 2.56 (s, 6H, αpy-C$\underline{H_3}$), 2.54 (s, 12H), 2.53 (s, 12H). ^{31}P NMR (162 MHz, acetone-d_6): δ/ppm, 21.6 (s, 4P, -$\underline{P}Ph_3$), – 143.7 (septet, 6P, $\underline{P}F_6^-$). FT-IR (in CH_2Cl_2) ν_{CO}/cm^{-1}: 1940, 1871. ESI–MS (in MeCN) *m/z*: 529 ([M-6PF_6^-]$^{6+}$), 663 ([M-5PF_6^-]$^{5+}$), 865 ([M-4PF_6^-]$^{4+}$). HRMS (ESI-TOF) *m/z*: [M-6PF_6^-]$^{6+}$ Calcd for $C_{170}H_{142}N_{16}O_4P_4Re_2Ru_2$ 528.6267; Found 528.6198, [M-5PF_6^-]$^{5+}$ Calcd for $C_{170}H_{142}F_6N_{16}O_4P_5Re_2Ru_2$ 663.3449; Found 663.3375.

Acknowledgements Figures 3.3, 3.5, and Scheme 3.2 were reproduced from the following references by permission of The Royal Society of Chemistry:

Y. Yamazaki, O. Ishitani, "Selectivity control between Mizoroki–Heck and homo-coupling reactions for synthesizing multinuclear metal complexes: unique addition effects of tertiary phosphines and O_2", *Dalton Trans.*, **2017**, *46*, 4816–4823.

References

1. Beletskaya IP, Cheprakov AV (2000) The heck reaction as a sharpening stone of palladium catalysis. Chem Rev 100:3009
2. Heck RF, Nolley JP (1972) Palladium-catalyzed vinylic hydrogen substitution reactions with aryl, benzyl, and styryl halides. J Org Chem 37:2320
3. Yamazaki Y, Morimoto T, Ishitani O (2015) Synthesis of novel photofunctional multinuclear complexes using a coupling reaction. Dalton Trans 44:11626
4. Lee K, Lee PH (2008) Efficient homo-coupling reactions of heterocyclic aromatic bromides catalyzed by $Pd(OAc)_2$ using indium.Tetrahedron Lett 49:4302
5. Ren T (2008) Peripheral covalent modification of inorganic and organometallic compounds through C–C bond formation reactions. Chem Rev 108:4185
6. Cassidy L, Horn S, Cleary L, Halpin Y, Browne WR, Vos JG (2009) Synthesis of asymmetric supramolecular compounds using a Ni(0) catalysed homo-coupling approach. Dalton Trans 3923
7. Takeda H, Yamamoto Y, Nishiura C, Ishitani O (2006) Analysis and isolation of cationic rhenium(I) and ruthenium(II) multinuclear complexes using size-exclusion chromatography. Anal Sci 22:545
8. Ozawa F, Kubo A, Hayashi T (1992) Generation of tertiary phosphine-coordinated Pd(0) species from $Pd(OAc)_2$ in the catalytic heck reaction. Chem Lett 2177
9. Reetz MT, de Vries JG (2004) Ligand-free Heck reactions using low Pd-loading. Chem Commun 1559
10. Furue M, Yoshidzumi T, Kinoshita S, Kushida T, Nozakura S-I, Kamachi M (1991) Intramolecular energy transfer in covalently linked polypyridine ruthenium(II)/osmium(II) binuclear complexes. Ru(II)$(bpy)_2$Mebpy-$(CH_2)_n$-MebpyOs(II)$(bpy)_2$ (n=2, 3, 5, and 7). Bull Chem Soc Jpn 64:1632

Chapter 4
Photochemical and Electrochemical Hydrogenation of π-Conjugated Bridging Ligands on Photofunctional Multinuclear Complexes

Abstract As described in Chaps. 2 and 3, a new class of multinuclear complexes, which have two photosensitizer units and a catalyst unit, was successfully synthesized by establishing the suitable reaction conditions for the Mizoroki–Heck reaction. However, the multinuclear complexes were linked through unsaturated vinylene linkers; therefore, they did not show high photocatalytic ability due to the lowered reducing power caused by the large π-conjugation. This chapter will report a photochemical hydrogenation method for converting the unsaturated linkers in the bridging ligands to their corresponding saturated carbon chains. The vinylene and ethynylene linkers in the bridging ligands of photofunctional multinuclear complexes synthesized by various coupling reactions were successfully converted to their corresponding saturated carbon chains under visible light irradiation in a MeCN−pyridine–CF_3COOH (3:1:0.1 v/v/v) mixed solution containing the starting metal complexes and a sacrificial electron donor in high yields. The new $Ru(II)_2$-Re(I) and Os(II)-Re(I)-Ru(II) trinuclear complexes, which showed high photocatalytic abilities for CO_2 reduction, were synthesized for the first time by combining the hydrogenation and the Mizoroki–Heck reactions.

Keywords Photochemical hydrogenation · Hetero-trinuclear complex · Photocatalytic CO_2 reduction · Electrochemical hydrogenation

4.1 Introduction

In Chaps. 2 and 3, I reported that the Mizoroki–Heck reaction and homo-coupling reaction were useful for coupling photofunctional metal complexes. In addition to these coupling reactions, in the reported literature, Sonogashira coupling [1, 2], olefin metathesis [3, 4], and Suzuki–Miyaura coupling [5–7], have also been used for connection of metal complexes possessing some functional groups on their ligands as active spots for the reactions, such as bromo groups, vinyl groups, ethynyl groups, and boronic acid substituents. By using these coupling reactions, various multinuclear complexes will be obtained to enlarge the variety of the multinuclear complexes.

Most of the complexes synthesized by the coupling methods possess unsaturated linkers, i.e., vinylene, ethynylene, and phenylene linkers. Such unsaturated

Y. Yamazaki, *Development of Synthetic Methods for Novel Photofunctional Multinuclear Complexes*, Springer Theses, https://doi.org/10.1007/978-981-16-7148-7_4

linkages between ligands have often led to an extension of π conjugation around the bridging ligands, which causes red-shift of absorption bands, increase of molar extinction coefficients, and emission lifetimes, as mentioned in Chaps. 2 and 3. These properties are useful when using these complexes as absorbers of dye-sensitized solar cells and/or photosensitizers of light-harvesting systems. However, the large π-conjugation has also caused a positive shift of reduction potentials of the obtained complexes. In the case of photocatalytic systems using multinuclear complexes as photocatalysts, such a decrease in reducing power of reduced forms of catalyst-complex units in multinuclear complexes should lower their reactivity with reagents, e.g., CO_2, and their photocatalytic abilities dramatically [8, 9]. Therefore, simple and easy synthetic strategies for converting the unsaturated linkers to the corresponding saturated carbon chains are desirable to increase the reducing power of multinuclear complexes synthesized by the coupling methods and to use them as efficient photocatalysts.

One of the most desirable reactions for the conversion of the unsaturated linkers should be hydrogenation reactions. However, most of the reported hydrogenation reactions are hardly applied to the olefins possessing bulky substituents because such olefins cannot approach the catalysts for the hydrogenation. The unsaturated linkers in the multinuclear complexes, surrounded by plural metal complexes, should have high steric hindrance; therefore, another approach should be developed to achieve the hydrogenation of the unsaturated linkers in multinuclear complexes synthesized using various coupling reactions.

Photocatalytic hydrogenation reactions of multi-substituted alkenes using photosensitizers and electron donors in protic solvents have already been reported [10]. In this mixture system, the alkenes were reduced by one-electron reduced species of the photosensitizer and then reacted with protons mainly derived from MeOH (Scheme 4.1). After further reduction and a reaction with protons, the hydrogenation reaction was accomplished. Multinuclear complexes synthesized by using coupling reactions mentioned in the previous chapters show photosensitizing abilities to form one-electron reduced species in the presence of electron donors under visible light irradiation. The electrons obtained photochemically should localize on the bridging ligands because of the positive redox potentials derived from the large π-conjugation. Therefore, the electron density of the vinylene linkers should become higher by photochemical reduction of the multinuclear complexes. In the presence of acids, these electron-rich vinylene linkers are expected to induce hydrogenation reactions (Scheme 4.2). The protons will approach the vinylene chains to trigger the hydrogenation reaction because protons are small ions, even though these vinylene chains have bulky substituents, i.e., plural metal-complex units.

In this chapter, I will report a photochemical hydrogenation reaction of vinylene chains (or ethynylene chains) in bridging diimine ligands on photofunctional multinuclear complexes by using their photosensitizing abilities. In this system, the photofunctional complexes could work not only as redox photosensitizers but also as reactants. The photochemically reduced species of the multinuclear complexes should induce the hydrogenation reaction via the reaction with protons and/or proton sources, which have low steric hindrance. In the presence of electron donors and

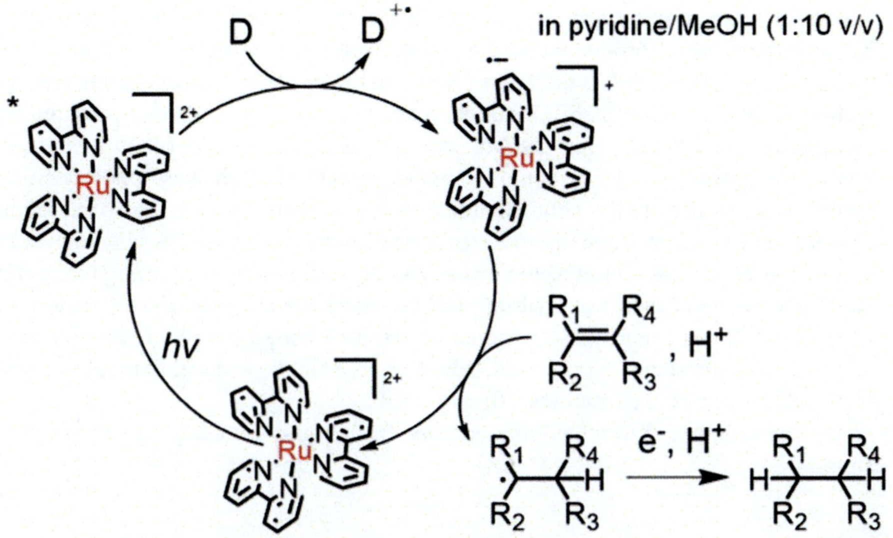

Scheme 4.1 Photochemical hydrogenation reaction of olefins using a redox photosensitizer ($[Ru(bpy)_3]^{2+}$) in the presence of electron donors

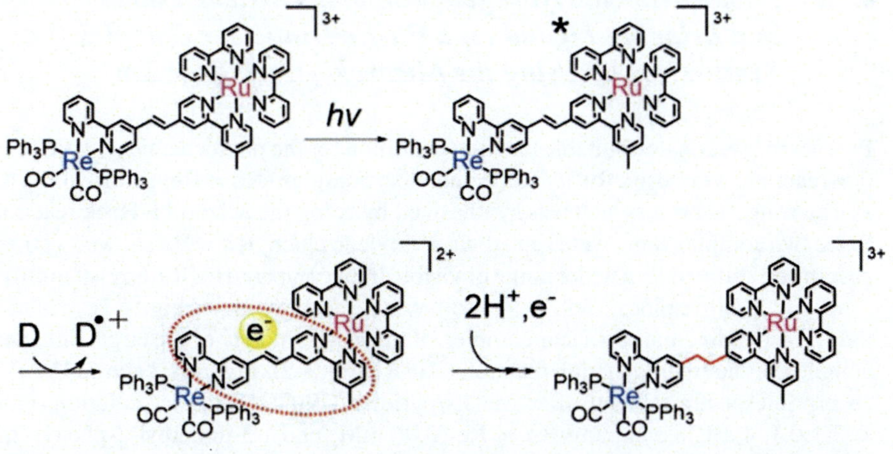

Scheme 4.2 Photochemical hydrogenation reaction of a vinylene linker in a π-conjugated bridging ligand of a photofunctional multinuclear complex synthesized via the Miziroki-Heck reaction using their photosensitizing abilities in the presence of electron donors and protons

proton sources, the hydrogenation reaction proceeded selectively under irradiation with relatively long-wavelength light, such as 600 nm. The starting materials possessing vinylene linkers could be selectively excited because of the red-shift of the absorption bands derived from the larger π-conjugation systems compared to the targeted products linked via saturated carbon chains. Hence, almost no byproducts derived from the photo-decomposition of products were observed after the reaction because the targeted products could not absorb the irradiation light. To the best of my knowledge, this result is the first example of the selective hydrogenation reaction using the photosensitizing abilities of the reactants themselves. Various multinuclear complexes synthesized using the Mizoroki–Heck reactions, the olefin metathesis, or the Sonogashira coupling could be hydrogenated through these photochemical methods. In addition to the photochemical hydrogenation reaction, I also succeeded in electrochemical hydrogenation. A multinuclear complex with a vinylene linker was dissolved in a MeCN-pyridine mixed solution containing supporting electrolyte and some acid and then was hydrogenated by bulk electrolysis. By combining the coupling methods using various coupling reactions with these hydrogenation reactions, novel multinuclear complexes, which showed high photocatalytic abilities for CO_2 reduction, could be successfully obtained.

In this chapter, different abbreviations from those in Chap. 2 are used for simplicity.

4.2 Results and Discussion

4.2.1 Photochemical Hydrogenation of a Vinylene Linker in a Bridging Ligand on a Ru-Re Dinuclear Complex Synthesized by Using the Mizoroki–Heck Reaction

In order to investigate a suitable reaction condition for the photochemical hydrogenation reaction, we chose a Ru(II)-Re(I) dinuclear complex (**Ru = Re**) shown in Eq. 4.1 as a starting material, which was synthesized by using the Mizoroki–Heck reaction. Since two complex units were linked via a vinylene chain, **Ru = Re** showed a strong absorption ability over a wider range of visible light compared to $[Ru(bpy)_2(^4dmb)]^{2+}$ (bpy = 2,2′-bipyridine, ^{4}dmb = 4,4′-dimethyl-2,2′-bipyridine, Fig. 4.1), which is the corresponding mononuclear complex of the Ru(II) moiety of the target dinuclear complex of the hydrogenation reaction (**Ru-Re**, Eq. 4.2). Figure 4.2 shows UV–Vis absorption spectra changes of an MeCN/pyridine/CF_3COOH mixed solution (v/v/v = 3:1:0.1, 4 mL) containing **Ru = Re** (0.06 mM) and 1,3-dimethyl-2-phenyl-2,3-dihydro-1*H*-benzo[*d*]imidazole (BIH, 50 mM) as an electron donor during irradiation with a 600 nm irradiation light (optical pass length: 1 cm, light intensity: 4.3×10^{-8} Einstein/s). In this reaction system, the excited state of **Ru = Re** was reductively quenched by BIH, and the reduced species of **Ru = Re** produced photochemically reacted with protons to convert the double bond to a single bond. Pyridine was used as a basic solvent that could remove protons from one-electron oxidized species of BIH to suppress back electron transfer from one-electron reduced species of **Ru = Re** to the radical cation of BIH [15]. CF_3COOH was used as a proton source. During the reaction, a blue-shift of the singlet metal-to-ligand charge transfer (^{1}MLCT) absorption band and a decrease in molar extinction coefficients were observed. The spectra

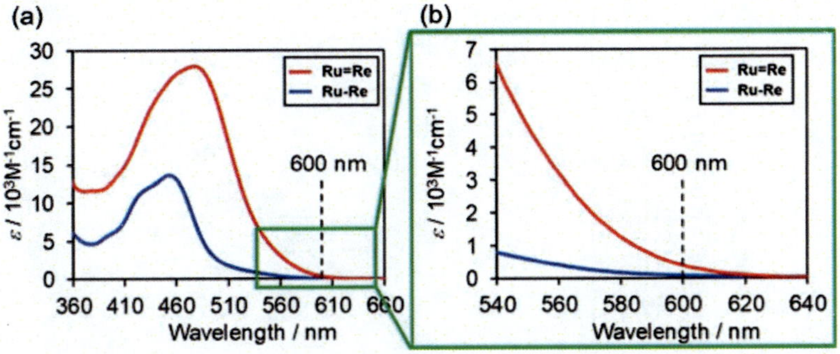

Fig. 4.1 Absorption spectra of **Ru = Re** (Red) and $[Ru(bpy)_2(dmb)]^{2+}$ (blue) measured in MeCN. The enlarged figure of (**a**) is shown in (**b**)

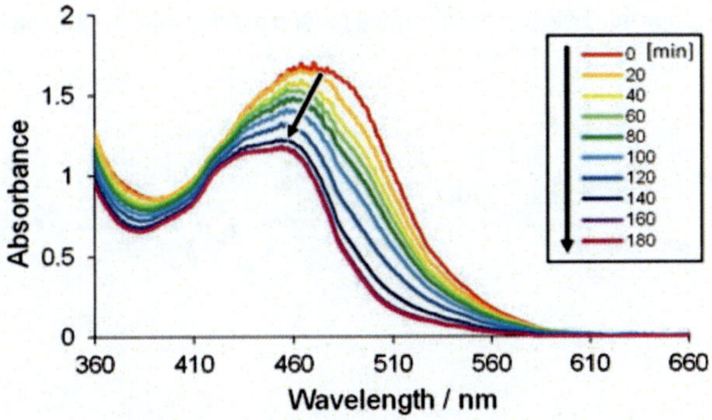

Fig. 4.2 Absorption spectra changes of the MeCN-pyridine mixed solution containing **Ru = Re**, CF_3COOH, and BIH during the irradiation with 600 nm excitation light (0–180 min)

changes were accomplished within approximately 4 h, and the shape of the spectrum became quite similar to that of $[Ru(bpy)_2(^4dmb)]^{2+}$ (Fig. 4.1). These indicated that all of the vinylene chains were converted to saturated carbon chains within 4 h. Further irradiation did not cause any spectral changes, which indicated that the products were not decomposed under the irradiation with 600 nm light. Figure 4.3 shows the relationship between the absorbed photon number and the decreased amount of **Ru = Re**, which is estimated from the differential absorbance detected at 600 nm. After an induction period, **Ru = Re** was linearly decreased with increasing the irradiation time. The quantum yield for this hydrogenation reaction was estimated at approximately 3% from the slope of the fitting curve.

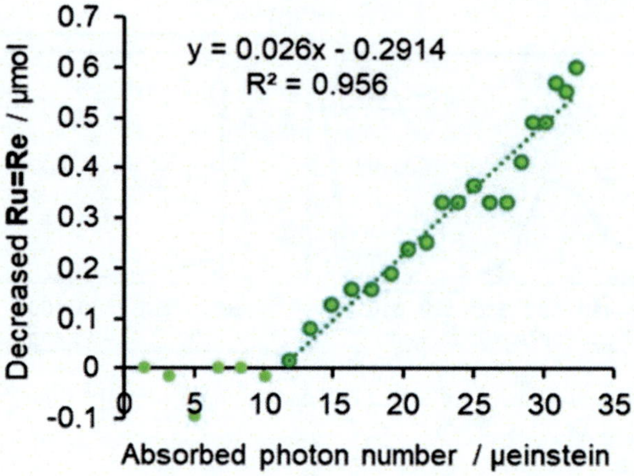

Fig. 4.3 The relationship between the absorbed photon number and the amount of **Ru = Re** reacted

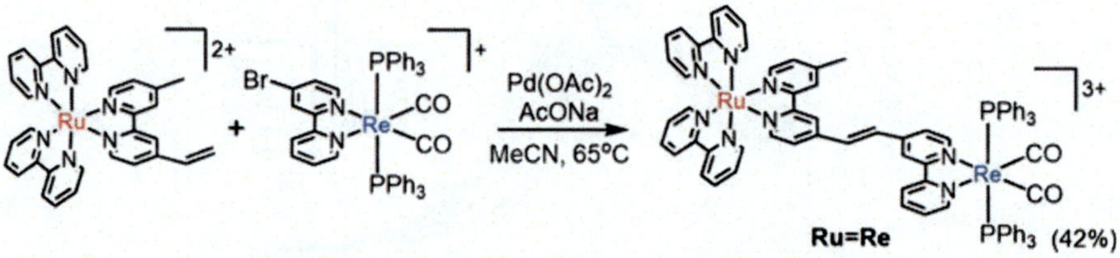

(4.1)

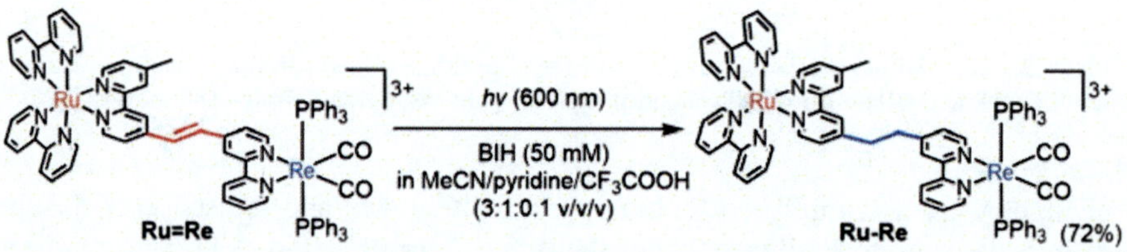

(4.2)

4.2.2 *Purification of the Products of Photochemical Hydrogenation of Ru = Re and Identification Using ESI-TOF–MS*

The crude products were purified in order to remove the electron donors and the oxidant of them by using ion-exchange chromatography [CM Sephadex, eluent: MeCN-water mixed solution (v/v = 1:1) containing NH_4PF_6 (0–8 mM)]. The product could be isolated with a relatively high yield (72%). The obtained product was analyzed by size exclusion chromatography. The size exclusion chromatograms

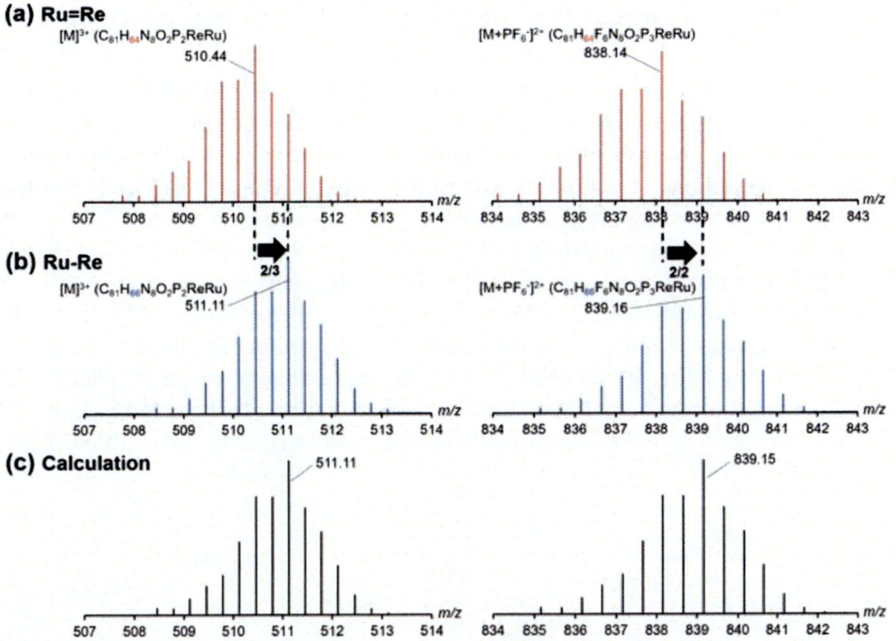

Fig. 4.4 ESI-Mass spectra of **a** **Ru = Re** and **b** **Ru-Re**. The calculated isotope model of **Ru-Re** is illustrated in (**c**)

measured before the reaction and after the purification were quite similar, which indicated that the products had almost the same molecular size as **Ru = Re**. The obtained products were also analyzed by electron spray ionization mass spectroscopy (ESI–MS, Fig. 4.4). The main peaks of the products at around 511.11 and 839.16 were shifted by $m/z = 2/3$ and 2/2 compared to those of **Ru = Re** ($m/z = 510.44$ and 838.14), respectively. The shape of the peaks was identical to the calculated isotope models for **Ru-Re**. These results clearly indicated that the photochemical hydrogenation reaction could be achieved with relatively high selectivity and isolation yield under the reaction condition mentioned above. A small amount of byproducts, probably derived from a side reaction between photochemically reduced species of **Ru = Re** and the oxidant of BIH, were observed in the third fraction of the ion-exchange chromatography in the spectrum of ESI–MS (first fraction: BIH and their oxidants, second fraction: **Ru-Re**, third fraction: the BIH adducts). The BIH adducts could be easily removed by the chromatography because they had different charges from that of **Ru-Re**.

4.2.3 Photochemical Hydrogenation with Different Solvents

The photochemical hydrogenation reactions were also conducted with other solvents, i.e., *N, N′*-dimethylformamide (DMF), and MeOH, instead of MeCN. In both cases, similar UV–Vis absorption spectra changes to the case using MeCN were observed during the reactions. Therefore, the vinylene linker in **Ru = Re** should have been converted to a corresponding saturated carbon chain. In the case using DMF, the spectral changes were accomplished within approximately 1 h. However, the ESI-mass spectrum of the main product was quite different from that with MeCN (Fig. 4.5). The main peaks were attributed to DMF-incorporated complexes. In the DMF solutions, it is supposed that not only protons but also DMF reacted with the one-electron reduced species of **Ru = Re**. When using MeOH, the main products were attributed to the hydrogenated target complex, though the reaction took place for more than 12 h. A small amount of MeOH-incorporated complexes were observed in the ESI-mass spectrum.

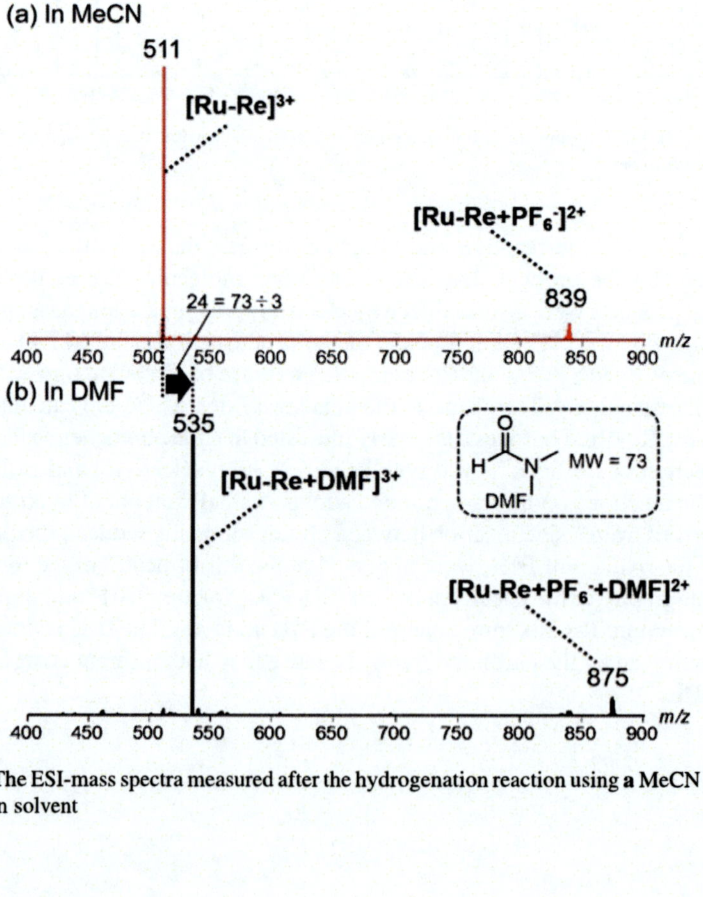

Fig. 4.5 The ESI-mass spectra measured after the hydrogenation reaction using **a** MeCN or **b** DMF as the main solvent

4.2.4 *Photochemical Hydrogenation of Various Dinuclear Complexes with a Vinylene Linker Synthesized by Using the Mizoroki–Heck Reaction*

As mentioned above, the photochemical hydrogenation could be achieved selectively in MeCN-pyridine mixed solutions containing BIH as an electron donor and CF_3COOH as a proton source with 600 nm irradiation light. This reaction could be applied to other photofunctional dinuclear complexes, which have different ligands and/or metal centers, i.e., Ir(III) (Eqs. 4.3 and 4.4). The isolation yields were summarized in Table 4.1. In the case of a Re(I) dinuclear complex (**Re(OEt) = Re(Et)**), which had different phosphine ligands on each unit, similar absorption change, i.e., blue-shift and decrease of absorption bands, was observed during the reactions, even though the reaction took place for more than 20 h probably due to small absorbance at 600 nm. The products were purified by ion-exchange chromatography and silica-gel column chromatography. The byproducts, i.e., BIH adducts, were obtained with a 29% isolation yield (Fig. 4.6). This is probably because **Re(OEt) = Re(Et)** has a relatively low steric hindrance around the vinylene linker compared to other complexes. The ligand cone angle of $P(OEt)_3$ and PEt_3 are 109° and 132°, respectively [11]. These values are smaller than PPh_3 (145°), which is contained in **Ru = Re**. Figure 4.7 shows the ^{1}H NMR spectra changes of the Re(I) dinuclear complex throughout the hydrogenation reaction from **Re(OEt) = Re(Et)** to **Re(OEt)-Re(Et)**. The peaks attributed to the olefinic protons on the bridging ligand in **Re(OEt) = Re(Et)** were disappeared, and new peaks derived from ethylene chains of **Re(OEt)-Re(Et)** were observed. When using an Ir-Re dinuclear complex (**Ir = Re**), obvious spectral changes could not be observed within 15 h under irradiation using 600 nm light. Therefore, the irradiation wavelength was changed to 510 nm, and then the absorption spectra changes were accomplished within 5 h.

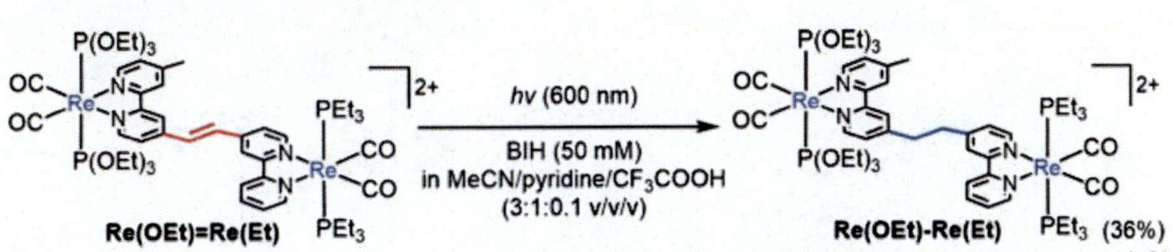

(4.3)

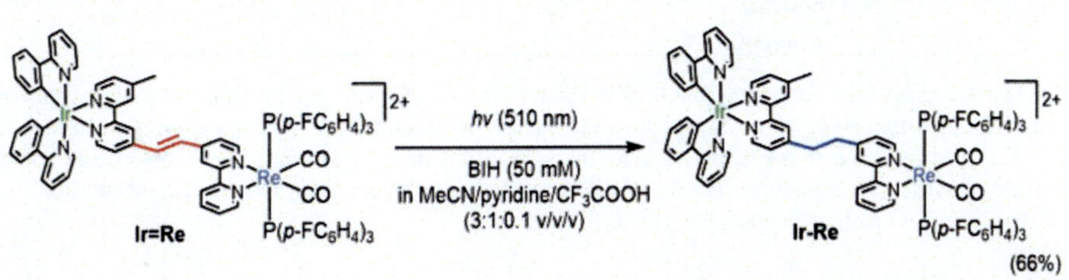

(4.4)

Table 4.1 Photochemical hydrogenation reactions of various photofunctional multinuclear complexes[a]

Entry	Substrates	Reaction time /h[b]	Products	Isolation Yield/%
1	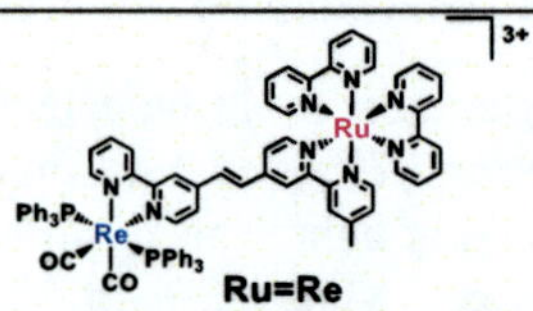 **Ru=Re**	6 (15)	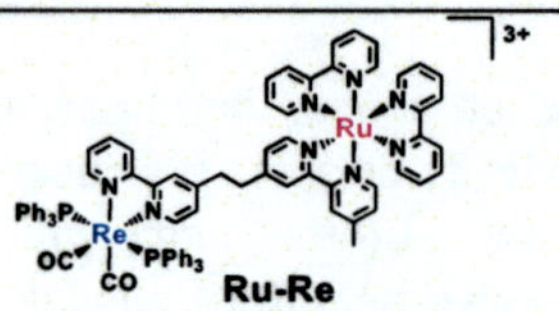 **Ru-Re**	72
2	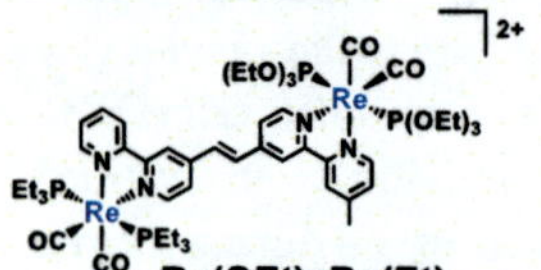 **Re(OEt)=Re(Et)**	21 (37)	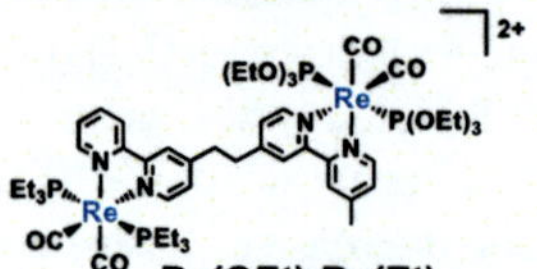 **Re(OEt)-Re(Et)**	36
3[c]	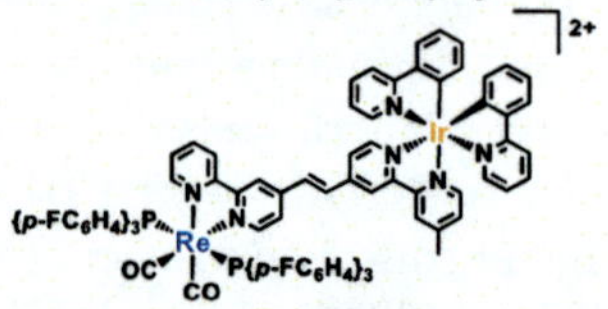 **Ir=Re**	5 (8)	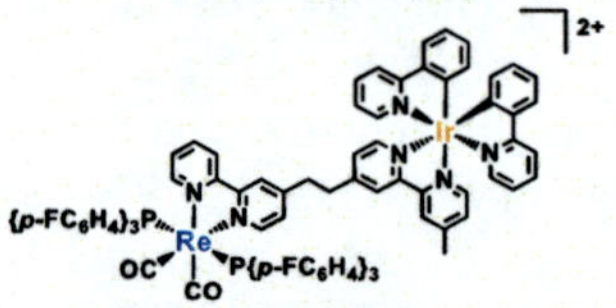 **Ir-Re**	66
4	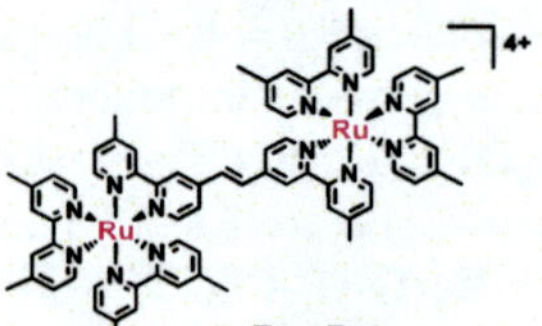 **Ru=Ru**	1 (15)	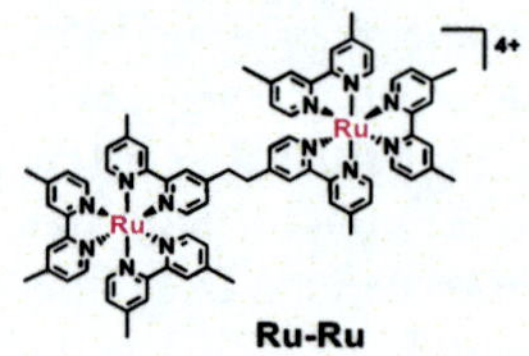 **Ru-Ru**	62
5	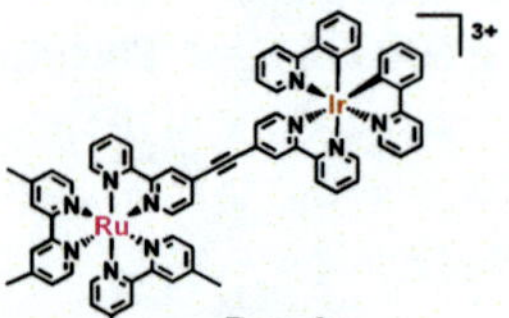 **Ru≡ Ir**	5 (14)	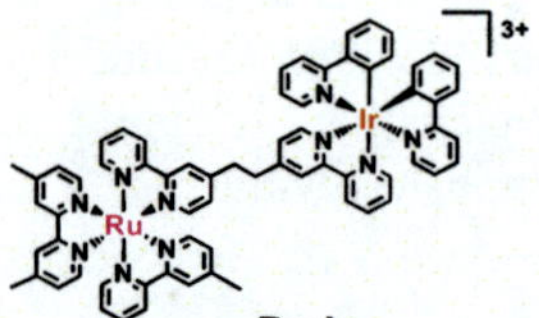 **Ru-Ir**	78
6	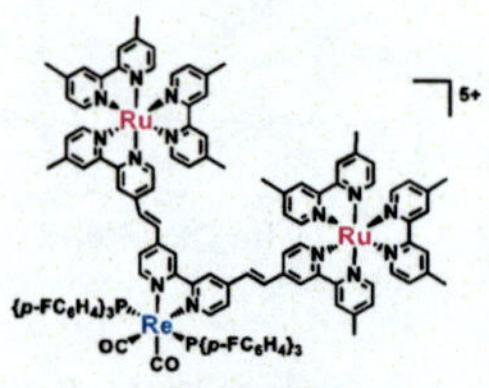 **Ru=Re=Ru**	7 (15)	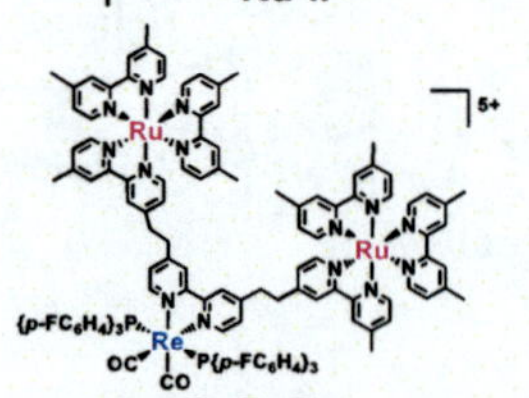 **Ru-Re-Ru**	55

[a] The reactions were conducted with MeCN-pyridine-CF_3COOH mixed solution [3:1:0.1 (v/v/v)] containing BIH (10 or 50 mM) and substrates under irradiation with 600 nm excitation light
[b] The reaction time was estimated from the irradiation time taken until when absorption spectra changes were accomplished. The values in parentheses indicate the practical irradiation time
[c] Irradiation wavelength was changed to 510 nm

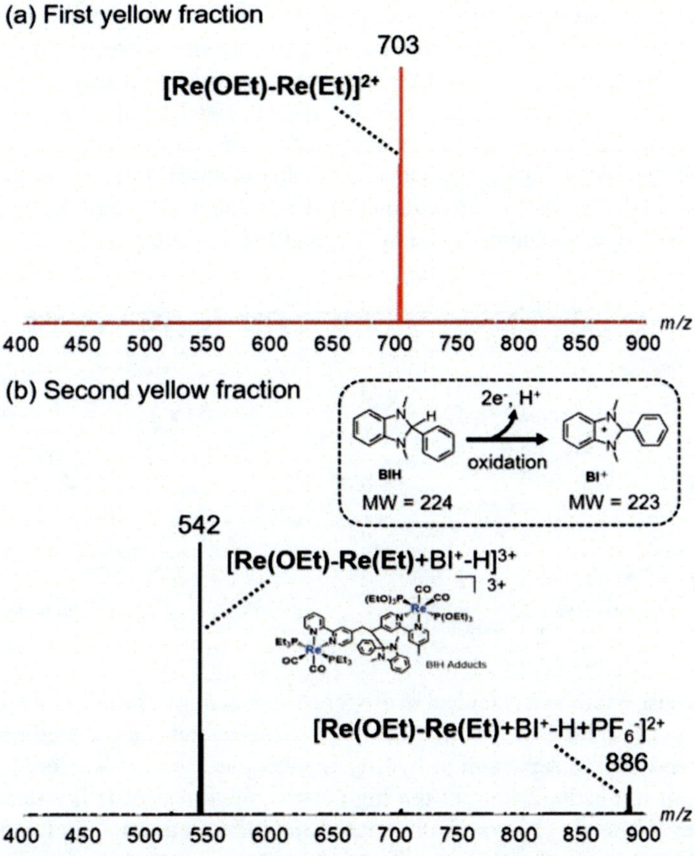

Fig. 4.6 The ESI-mass spectra of the products contained in **a** the first yellow fraction and **b** the second yellow fraction of the ion-exchange chromatography conducted after the hydrogenation reaction of **Re(OEt) = Re(Et)**

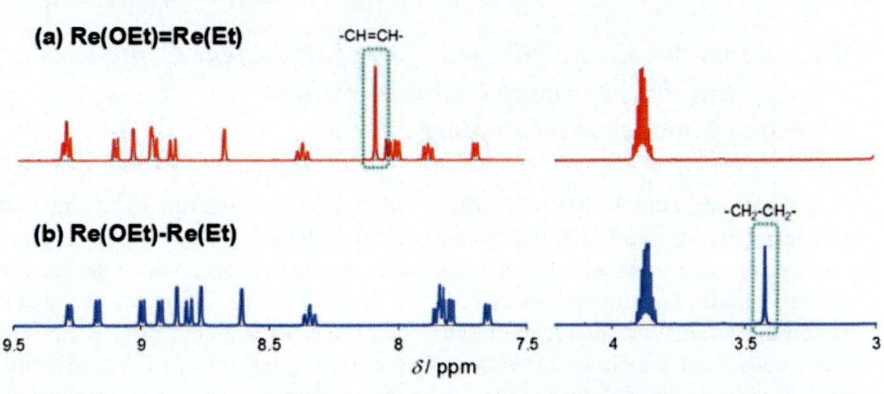

Fig. 4.7 ^{1}H NMR spectra of **a Re(OEt) = Re(Et)** and **b Re(OEt)-Re(Et)** in acetone-d_6

The hydrogenation reaction could be applied for not only dinuclear complexes but also trinuclear complexes, which have two vinylene linkers (**Ru = Re = Ru**, Eq. 4.5). The absorbance at 600 nm decreased linearly with increasing irradiation time, and both vinylene linkers could be hydrogenated within 7 h under the same reaction condition mentioned above. The purification could be achieved by using ion-exchange chromatography, and isolation yields reached 55%. From these results, it is supposed that various photofunctional multinuclear complexes linked by π-conjugated vinylene chains can be hydrogenated photochemically.

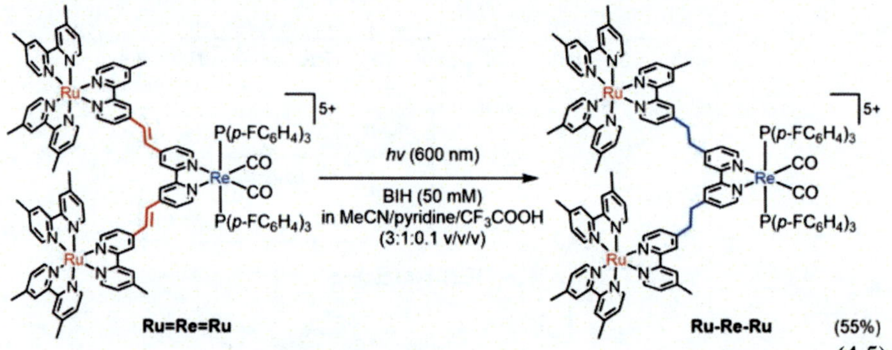

(4.5)

I have also examined chemical hydrogenation reactions of Ru(II)$_2$-Re(I) trinuclear complexes using Pd/C or Crabtree catalysts. Unfortunately, however, the reaction did not proceed well under 1 atm of hydrogen atmosphere and at ambient temperature. This result might indicate that the high steric hindrance of bulky metal-complex units prevents those catalysts from approaching the olefin moiety. It is noteworthy that the photochemical hydrogenation mentioned above is useful even when the olefin possesses bulky substituents.

4.2.5 *Photochemical Hydrogenation of Dinuclear Complexes Synthesized by Using Olefin Metathesis and Sonogashira Coupling*

A Ru(II) dimer complex (**Ru = Ru**), which was synthesized by using olefin metathesis (Eq. 4.6), and a Ru-Ir dinuclear complex (**Ru≡Ir**), which was synthesized by using Sonogashira coupling (Eq. 4.7), could also be hydrogenated under the same reaction condition mentioned above (Eqs. 4.8 and 4.9). This result indicates that not only double bonds but also triple bonds could be hydrogenated selectively. From these results, I can conclude that various C–C coupling reactions, i.e., the Mizoroki–Heck reaction, olefin metathesis, and Sonogashira coupling, could be combined with the photochemical hydrogenation for developing novel photofunctional complexes and tuning their photophysical properties and/or photocatalytic abilities.

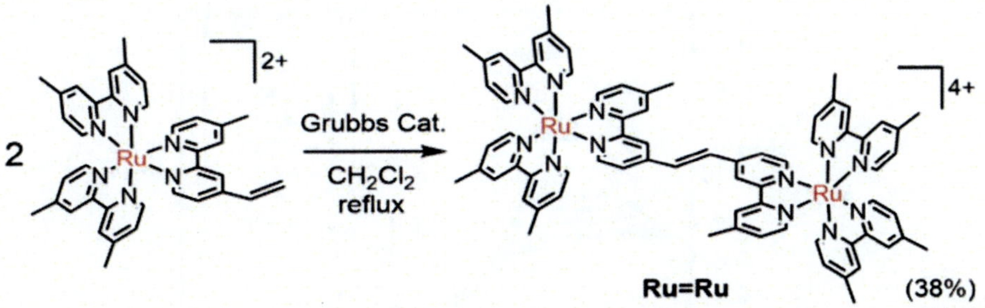
2
Grubbs Cat.
CH2Cl2
reflux
Ru=Ru
(38%)

(4.6)

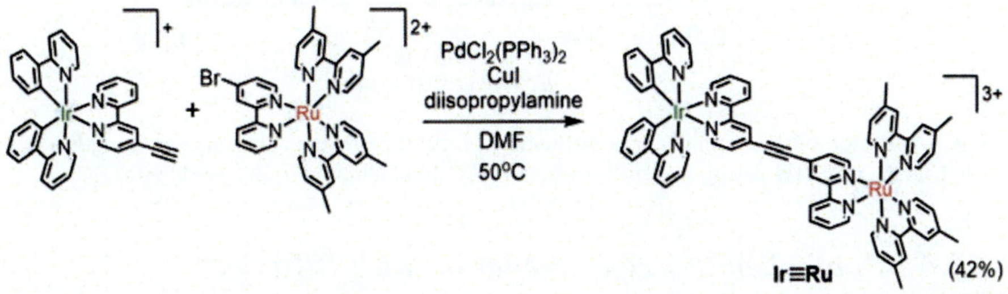
PdCl2(PPh3)2
CuI
diisopropylamine
DMF
50°C
Ir≡Ru
(42%)

(4.7)

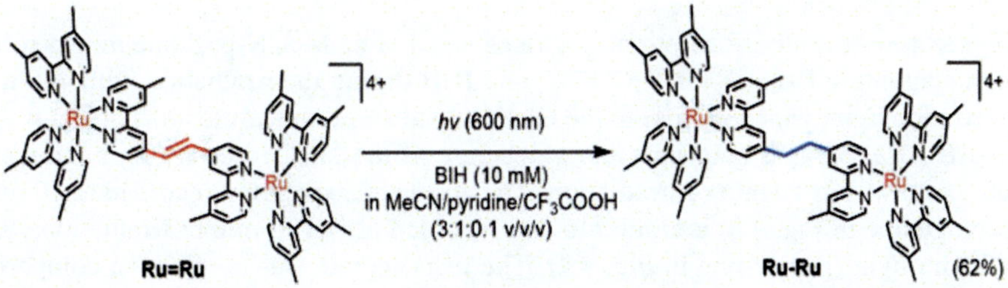
hv (600 nm)
BIH (10 mM)
in MeCN/pyridine/CF3COOH
(3:1:0.1 v/v/v)
Ru=Ru
Ru-Ru
(62%)

(4.8)

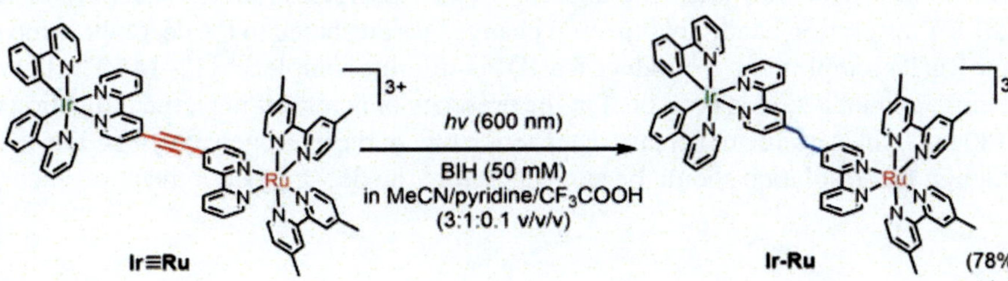
hv (600 nm)
BIH (50 mM)
in MeCN/pyridine/CF3COOH
(3:1:0.1 v/v/v)
Ir≡Ru
Ir-Ru
(78%)

(4.9)

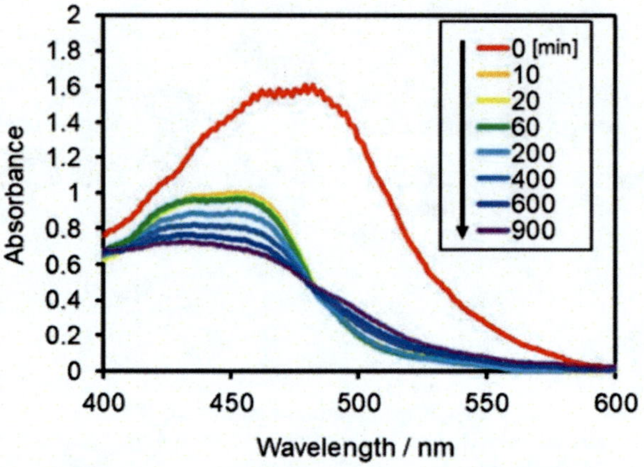

Fig. 4.8 Absorption spectra changes of the MeCN-pyridine mixed solution containing **Ru = Re**, CF_3COOH, and BIH during the irradiation with 480 nm excitation light (0–900 min)

4.2.6 Photochemical Hydrogenation with Different Wavelength Irradiation Light

Figure 4.8 shows the absorption spectra changes of an MeCN-pyridine mixed solution containing **Ru = Re**, CF_3COOH, and BIH during the irradiation with 480 nm excitation light, which is close to the maximum absorption wavelengths of **Ru = Re** and **Ru-Re**. A rapid blue shift and a decrease of the MLCT absorption band were observed within 10 min of irradiation. The absorption spectrum recorded at 10 min (orange line in Fig. 4.8) is similar to that recorded after 180-min of irradiation with 600 nm light (purple line in Fig. 4.2). The reaction rate was accelerated compared to the case using 600 nm light because of the much larger absorptivity of **Ru = Re** at 480 nm. However, further irradiation caused decreasing and broadening of the MLCT absorption band. This spectral change was attributed to the decomposition of the Ru(II) complex unit to produce Ru(II) bis-diimine complexes [12–14]. The ligand dissociation reaction is promoted in the presence of acids. Of note, the utilization of 480 nm light is useful to increase the reaction rate of the hydrogenation reaction, even though the irradiation should be stopped before the decomposition reaction occurs.

4.2.7 Electrochemical Hydrogenation of the Ru-Re Dinuclear Complex with a Vinylene Linker

The hydrogenation reaction was achieved by not only photochemical processes mentioned above but also electrochemical processes. Figure 4.9 shows a cyclic voltammogram of **Ru = Re** (0.1 mM) measured in a MeCN solution containing NEt_4BF_4 (0.1 M) as a supporting electrolyte. In the voltammogram, two reversible

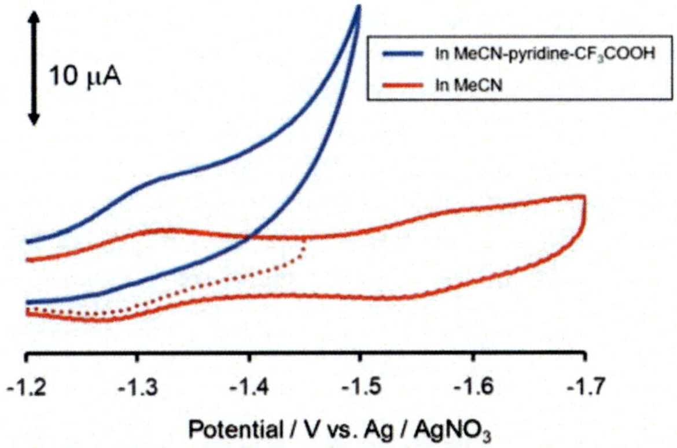

Fig. 4.9 Cyclic voltammograms of **Ru = Re** (0.1 mM) measured in MeCN (red line) and MeCN/pyridine/CF_3COOH (3/1/0.1 v/v/v, blue line) containing NEt_4BF_4 (0.1 M) under Ar atmosphere. The sweep scan rate was 200 mV/s

reduction waves, which were attributed to the stepwise reduction of the bridging ligand, were observed. When adding CF_3COOH as a proton source and pyridine as a base, which could neutralize the acid and suppress efficient hydrogen evolution, the first reduction wave became irreversible. This result strongly suggests that the one-electron reduced species of **Ru = Re** could react with protons. Therefore, it is expected that one-electron reduced species of **Ru = Re** produced electrochemically should also react with protons. The bulk electrolysis was conducted in a MeCN/pyridine/CF_3COOH mixed solution (3:1:0.1 v/v/v) containing Et_4NBF_4 (0.1 M) as a supporting electrolyte with mesh-texture glassy-carbon working electrode, an Ag/$AgNO_3$ (0.01 M) reference electrode, and a mesh-texture Pt counter electrode. The applied potential was − 1.5 V versus Ag/$AgNO_3$. The bulk electrolysis caused a color change from red to orange (Fig. 4.10). The current gradually decreased

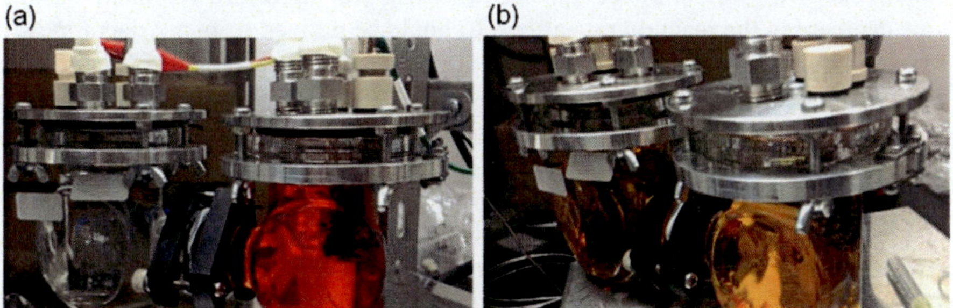

Fig. 4.10 The reaction solution of **Ru = Re** (0.1 mM) [MeCN/pyridine/CF_3COOH (3/1/0.1 v/v/v) containing NEt_4BF_4 (0.1 M)] **a** before and **b** after bulk electrolysis

over 7 h of the bulk electrolysis. After 10 h of the bulk electrolysis, the product was purified by diluting with CH_2Cl_2 for washing with water and ion-exchange chromatography (CM Sephadex C-25, eluent: MeCN-water (1:1 v/v) containing NH_4PF_6 (0–8 mM)). The isolation yield was 59%. The faradaic efficiency of this reaction was not high (5%), probably because of the competing hydrogen evolution. These results demonstrate that multinuclear complexes linked through unsaturated carbon chains synthesized using coupling reactions can be hydrogenated by the electrochemical hydrogenation reaction, even when the complexes do not show photofunctional properties.

4.2.8 Photocatalytic Ability of Ru-Re-Ru for CO_2 Reduction

Ru-Re-Ru was used for photocatalytic CO_2 reduction to investigate the effect of the hydrogenation reaction on photocatalytic abilities (Fig. 4.11). A mixture of DMF and triethanolamine (5:1 v/v) containing 0.05 mM of **Ru-Re-Ru** and 0.1 M of 1-benzyl-1,4-dihydronicotinamide (BNAH) was irradiated using a high-pressure mercury lamp equipped with a solution filter ($\lambda_{ex} > 500$ nm). The photocatalysis produced CO as the main product and formate and hydrogen as minor products. After 20 h of irradiation, the turnover number (TON) of CO, formate, and H_2 reached 313, 85, and 22, respectively. The TON_{CO} and selectivity for the CO production with **Ru-Re-Ru** were higher than those with **Ru = Re = Ru** as a reference ($TON_{CO} = 283$, $TON_{HCOOH} = 100$, $TON_{H2} = 26$, Fig. 4.11). Of note, the reaction rate of CO production with **Ru-Re-Ru** was higher than that with **Ru = Re = Ru**. Besides, a longer induction period before the formate production was observed when using **Ru-Re-Ru** compared with when using **Ru = Re = Ru**. In the SEC charts of the reaction solutions containing **Ru-Re-Ru** recorded after the photocatalytic reaction, obvious peaks were detected around 33 min (Fig. 4.12a). This means that the reaction solutions, even after 20 h of irradiation, contained the starting complex and/or ligand-substituted complexes with similar molecular size to **Ru-Re-Ru**. Therefore, the main reason for the decrease in the reaction rate after 5 h should be derived from not decomposition of the complexes, but the accumulation of BNA_2, which is the dimerized compound of oxidants of BNAH, as previously reported (Eq. 4.10) [15, 16]. BNA_2 is known as an inhibitory compound of photocatalytic CO_2 reduction because $BNA_2^{\bullet+}$ produced after the reductive quenching process is stable and can oxidize one-electron reduced species of photosensitizers to recover BNA_2 (Eq. 4.11). These processes diminish the electrons for the catalytic reaction to waste the absorbed photons. In contrast, in the case of **Ru = Re = Ru**, significant peaks were observed around 36 min in the SEC chart recorded only after 1 h of irradiation (Fig. 4.12b). From ESI-mass spectroscopy, these peaks were attributable to the complexes, of which the Ru unit was decomposed to be bis-diimine type complex. This decomposition of the Ru units should be the reason for the larger TON_{HCOOH} and the shorter induction period before the formate production because the bis-diimine complexes are well-known catalysts for formate production. These results clearly show that both the reaction

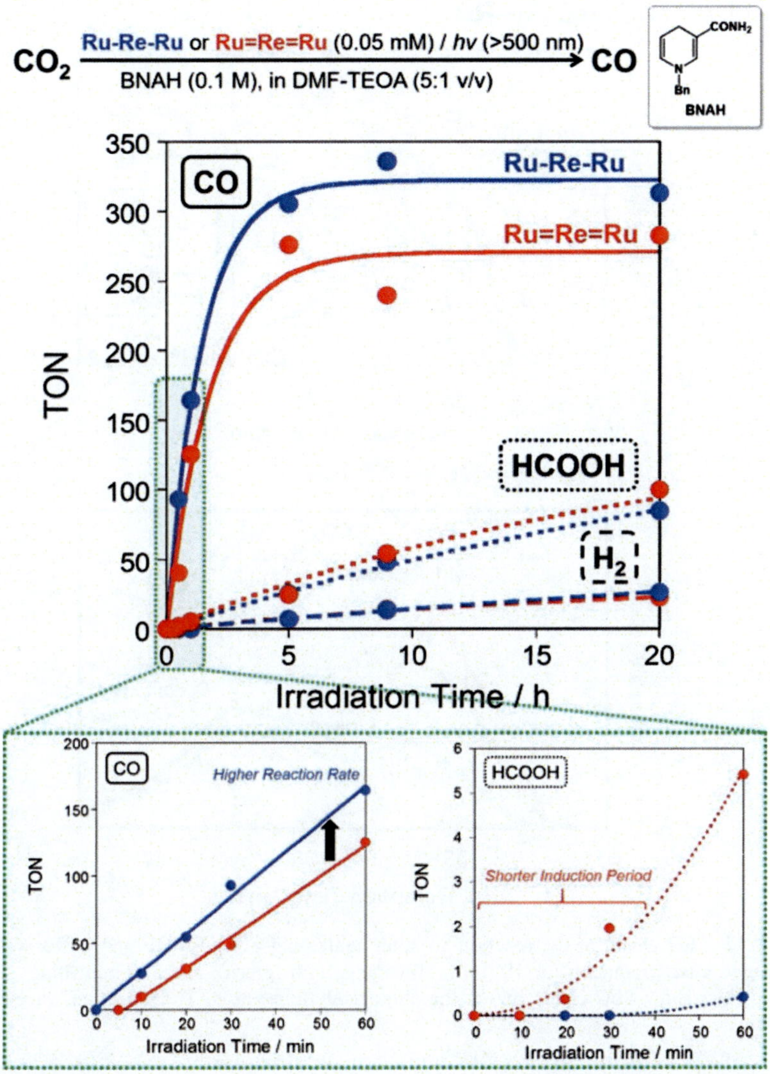

Fig. 4.11 Time courses of the TON of reduction products, i.e., CO (solid line), HCOOH (dotted line), and H_2 (broken line), during the photocatalytic reaction (20 h, $\lambda_{ex} > 500$ nm) using a mixture of DMF and triethanolamine (5:1 v/v) containing and 0.1 M BNAH, and 0.05 mM **Ru = Re = Ru** (red) or **Ru-Re-Ru** (blue)

selectivity and the durability of the multinuclear complexes were improved by the hydrogenation reaction.

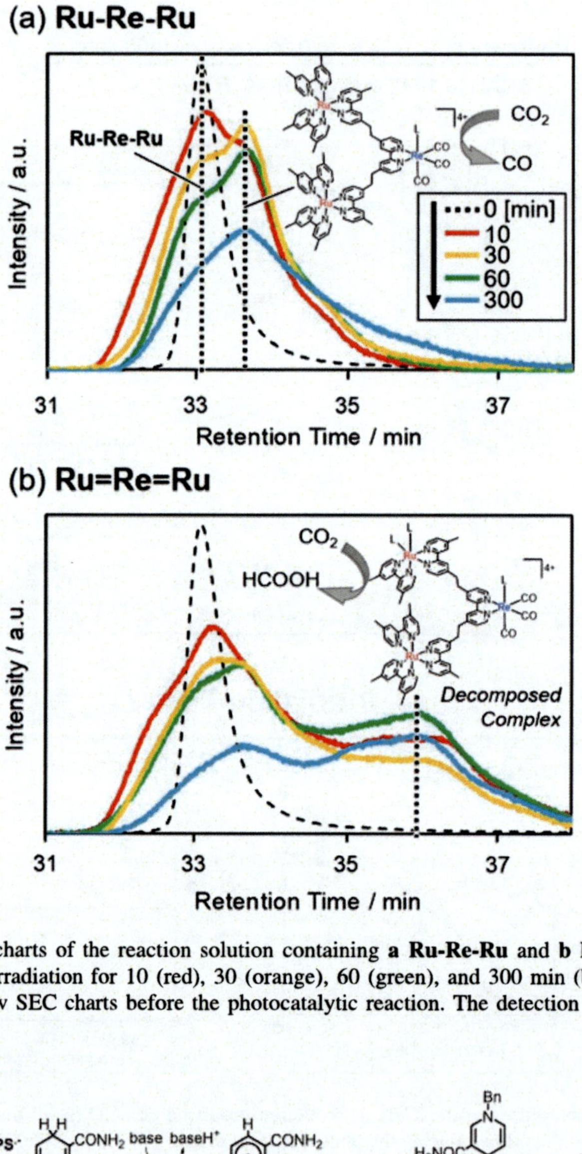

Fig. 4.12 SEC charts of the reaction solution containing **a Ru-Re-Ru** and **b Ru = Re = Ru** measured after irradiation for 10 (red), 30 (orange), 60 (green), and 300 min (blue). The black dotted lines show SEC charts before the photocatalytic reaction. The detection wavelength was 420 nm

(4.10)

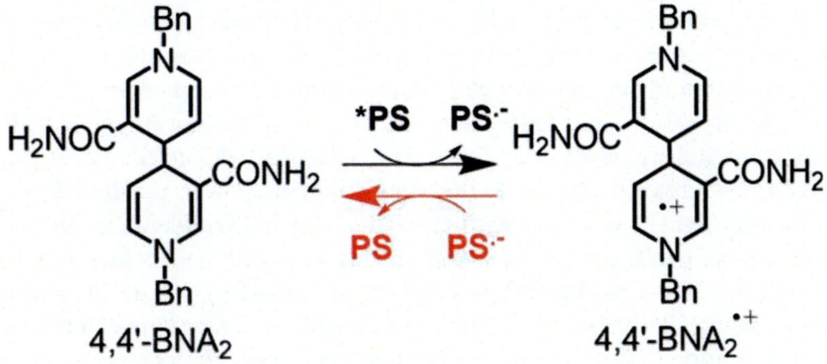

(4.11)

4.2.9 *Development of an Efficient Photocatalyst by Using Both Stepwise Mizoroki–Heck Reactions and the Photochemical Hydrogenation Reaction*

The hydrogenation reaction could be applied for various multinuclear complexes synthesized using coupling reactions, and their photocatalytic abilities for CO_2 reduction could be improved by the hydrogenation reaction as described above. Therefore, some new and efficient multinuclear photocatalysts were expected to be synthesized by a combination of the hydrogenation reaction and stepwise Mizoroki–Heck reactions mentioned in Chap. 3, which could connect two kinds of metal complexes with one Re(I) catalyst complex. I aimed to synthesize a trinuclear complex **Os-Re-Ru** shown in Fig. 4.13, which had two kinds of useful photosensitizer complexes, i.e., a Ru tris-diimine complex and an Os tris-diimine complex. Photocatalytic abilities

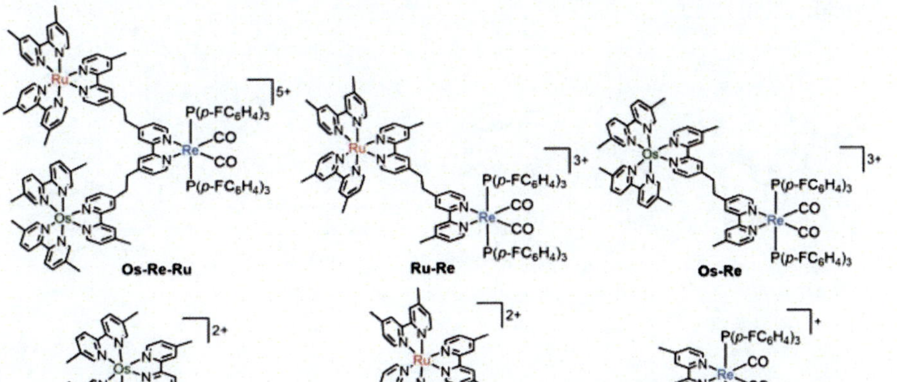

Fig. 4.13 Structures of **Os-Re-Ru** and the corresponding model complexes

of the dinuclear complexes, **Ru-Re** [17] and **Os-Re** [18], have already been investigated. In the case of **Os-Re**, the photocatalytic reaction could take place even under red-light irradiation owing to the triplet MLCT absorption bands between 540 and 730 nm, even though the excited state of the Os unit could not be quenched efficiently even by the strong electron donor, i.e., 1,3-dimethyl-2-phenyl-2,3-dihydro-1*H*-benzo[*d*]imidazole (BIH), due to the weak oxidizing power. On the other hand, **Ru-Re** showed higher photocatalytic ability than that of **Os-Re** under irradiation with short-wavelength light such as 480 nm owing to the high quenching fraction of the Ru unit (quantum yield for CO production: Φ_{CO}(**Ru-Re**) = 0.45, Φ_{CO}(**Os-Re**) = 0.12). If both the Os and Ru photosensitizer complexes can be connected with one Re(I) catalyst complex, the multinuclear complex can trigger photocatalytic reactions under irradiation of a wide range of visible light with high efficiency. Even though the synthesis of such hetero-trinuclear complexes is difficult using the conventional synthetic method with bridging ligands, they will be synthesized using a combination of the stepwise Heck reactions described in Chap. 3 and the hydrogenation reaction described in this chapter. To the best of my knowledge, this complex is the first example of a photocatalyst possessing two kinds of photosensitizers.

Equations 4.12, 4.13, and 4.14 show the synthetic schemes of the Os(II)-Re(I)-Ru(II) trinuclear complex (**Os-Re-Ru**). Using two types of reaction conditions for the Mizoroki–Heck reaction, i.e., the reaction in the presence of phosphine ligands with air injection and that in the absence of phosphine ligands with an excess amount of complexes having vinyl groups, the trinuclear complex linked via vinylene chains (**Os = Re = Ru**) was obtained with a moderate yield as well as the reactions described in Chap. 3. **Os = Re = Ru** was successfully converted to **Os-Re-Ru** by irradiating with 510 nm light in the presence of electron donors (BIH) and proton sources (CF_3COOH). This wavelength can mainly excite **Os = Re = Ru** because of the red-shift of the MLCT absorption bands compared to that of **Os-Re-Ru**. The absorption spectra changes were accomplished within approximately 10 h. After the purification using ion-exchange chromatography, **Os-Re-Ru** was obtained with a 52% of isolation yield.

(4.12)

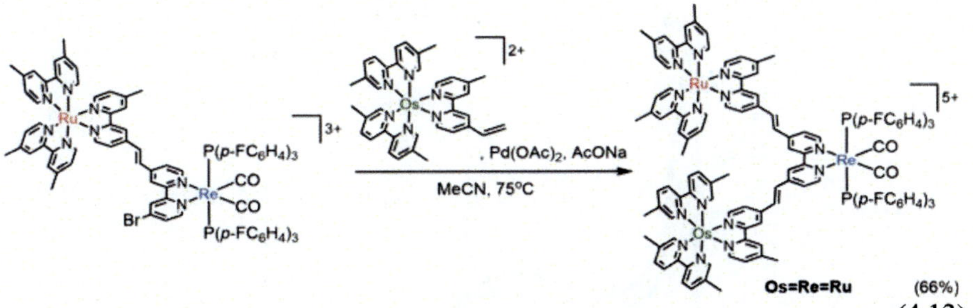

(4.13)

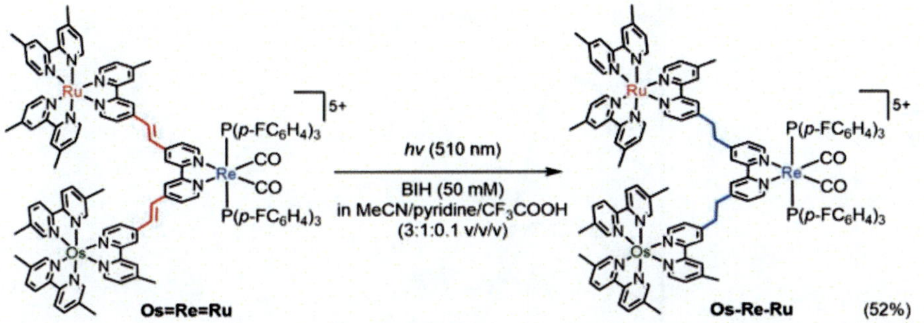

(4.14)

Figure 4.14 shows UV–Vis absorption spectra of **Os-Re-Ru** and the corresponding mononuclear complexes, i.e., $[Os(^5dmb)_2(^4dmb)]^{2+}$ (**Os**, ^{5}dmb = 5,5′-dimethyl-2,2′-bipyridine, ^{4}dmb = 4,4′-dimethyl-2,2′-bipyridine), $[Re(^4dmb)(CO)_2\{P(p\text{-}FC_6H_4)_3\}_2]^+$ (**Re**) and $[Ru(^4dmb)_3]^{2+}$ (**Ru**). The spectrum of **Os-Re-Ru** is almost identical to the 1:1:1 summation spectrum of **Os**, **Re**, and **Ru**. By introducing two kinds of photosensitizer complexes, i.e., **Os** and **Ru**, **Os-Re-Ru** shows strong absorption ability over a wide range of visible regions (~730 nm). Therefore, **Os-Re-Ru** can absorb strongly over a wide range of visible light compared with **Ru-Re** and **Os-Re**.

Next, we investigated the photocatalytic ability of **Os-Re-Ru** by comparing it with those of **Os-Re** and **Ru-Re**. A mixture of *N,N*-dimethylacetamide (DMA) and triethanolamine (TEOA) (5:1 v/v) containing 0.01 mM **Os-Re-Ru** and 0.2 M BIH as a sacrificial electron donor was irradiated using a halogen lamp equipped with a K_2CrO_4 (30%, w/w, $d = 1$ cm) solution filter ($\lambda_{ex} > 480$ nm). Highly selective CO production was observed with very small amounts of hydrogen and formate (TON_{H2}, $TON_{HCOOH} < 1$). The time courses of CO production using **Os-Re-Ru**, **Os-Re**, or **Ru-Re** are illustrated in Fig. 4.15. TON_{CO} reached 3550 ± 460 after 35 h of irradiation. This TON_{CO} is higher than those using **Ru-Re** ($TON_{CO} = 2810$) or **Os-Re** ($TON_{CO} = 850$), under the same reaction conditions. Because of the efficient photosensitizing reaction by the Ru unit, **Os-Re-Ru** showed higher photocatalytic ability than **Os-Re**. **Os-Re-Ru** also showed higher durability than **Ru-Re**. However, the reaction rate in the initial stage of the photocatalytic reaction was similar to each other, even though **Os-Re-Ru** can absorb more photons than **Ru-Re** (Fig. 4.14).

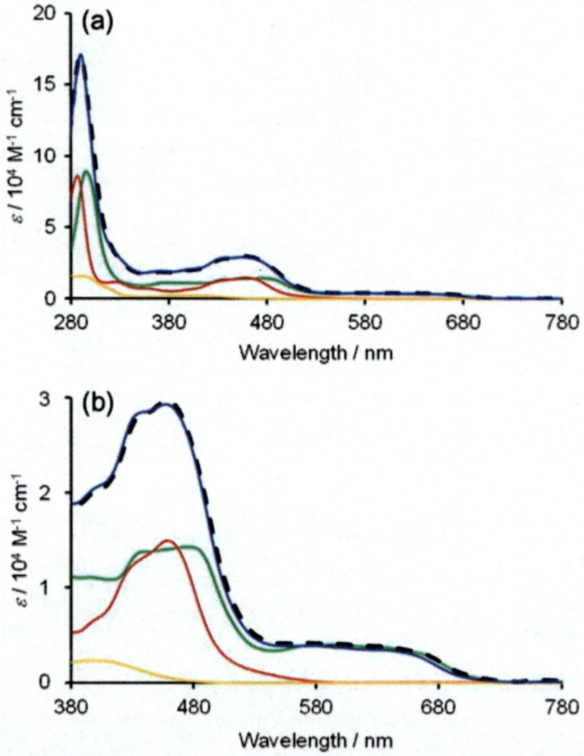

Fig. 4.14 **a** UV–Vis absorption spectra of the obtained trinuclear complex **Os-Re-Ru** (blue) and corresponding mononuclear complexes, i.e., **Os** (green), **Re** (yellow), and **Ru** (red). The 1:1:1 summation spectrum of **Os**, **Re,** and **Ru** are illustrated as a black dotted line. The solvent was MeCN. The enlarged figure of (**a**) (380–780 nm) is shown in (**b**)

This result might be caused by intramolecular energy transfer from the Ru unit to the Os unit. Figure 4.16 shows the emission spectra of **Os-Re-Ru** recorded with various excitation wavelengths. The spectra have two bands, of which maxima were detected at around 650 and 750 nm. These two bands were attributable to the emission from the Ru unit and the Os unit, respectively. Even though the emission quantum yield of the corresponding mononuclear complexes of the Ru unit (Φ_{em}(**Ru**) = 0.089) [17] is much higher than that of **Os** (Φ_{em}(**Os**) = 0.007) [18], the observed spectra were constructed mainly with the emission band of the Os unit regardless of the excitation wavelength. Therefore, the excitation energy of the Ru unit should partially be transferred to the Os unit. It has been reported that the quenching fraction of **Os** (43%) [18] is lower than that of **Ru** (99%) [17]; therefore, the energy transfer from the Ru unit to the Os unit might lower the quenching fractions and the reaction rate of **Os-Re-Ru**.

Figure 4.17 illustrates the difference absorption spectrum of a DMA-TEOA mixed solution (5:1 v/v) containing 0.05 mM **Os-Re-Ru** and 0.1 M BIH recorded after

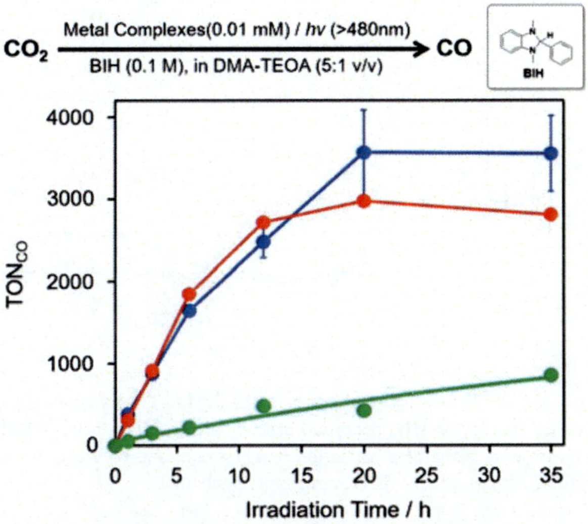

Fig. 4.15 Time courses of the TON of CO formation during photocatalytic reactions (35 h, $\lambda_{ex} >$ 480 nm) using a mixture of DMA and triethanolamine (5:1 v/v) containing 0.1 M BIH and 0.01 mM metal complexes (blue: **Os-Re-Ru**, red: **Ru-Re**, green: **Os-Re**)

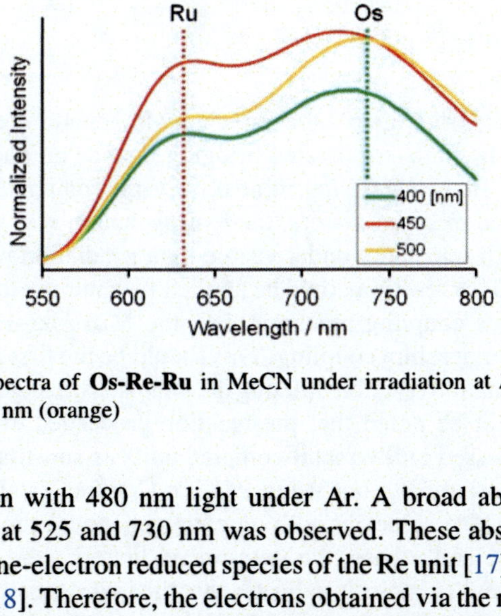

Fig. 4.16 Emission spectra of **Os-Re-Ru** in MeCN under irradiation at λ_{ex} = 400 nm (green), 450 nm (red), and 500 nm (orange)

1 min of irradiation with 480 nm light under Ar. A broad absorption band with two obvious peaks at 525 and 730 nm was observed. These absorption bands were attributable to the one-electron reduced species of the Re unit [17] and those of the Os unit, respectively [18]. Therefore, the electrons obtained via the reductive quenching processes localized on the Re and Os units. Of note, the difference spectrum hardly contained the absorption band derived from the one-electron reduced species of the Ru unit, even though both the Os and Ru units absorbed the irradiation light; the electrons obtained by the Ru units were transferred to the other units. This should be the reason for the higher durability of the Ru unit in **Os-Re-Ru** and the higher TON_{CO}

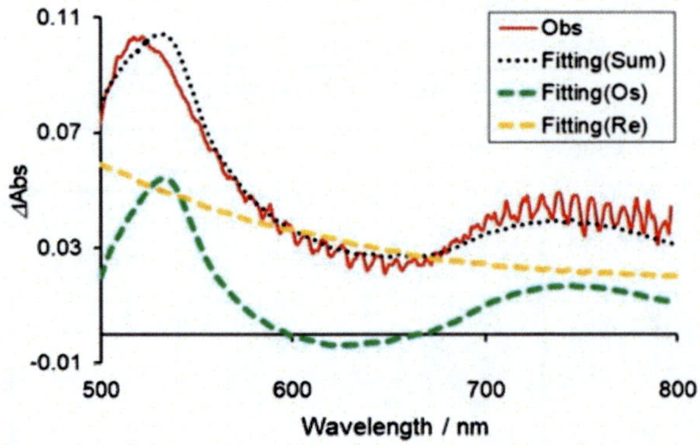

Fig. 4.17 The difference absorption spectrum of the DMA-TEOA (5:1 v/v) solution containing 0.05 mM **Os-Re-Ru** and 0.1 M BIH recorded after 1 min irradiation with 480 nm light under Ar. The simulation spectra are illustrated as dotted lines (green: one-electron reduced species (OERS) of **Os**, orange: OERS of **Re**, black: the summation spectrum)

because one-electron reduced species of Ru tris-diimine complexes are decomposed by absorbing the irradiation light in the reaction solutions, as described above.

4.3 Conclusion

In this chapter, I reported the photochemical hydrogenation reaction of vinylene or ethnylene linkers in the π-conjugated bridging ligands on photofunctional multinuclear complexes. In the MeCN/pyridine mixed solution containing the complexes, proton sources, and electron donors, the hydrogenation was successfully achieved with relatively high selectivity under visible light irradiation with long wavelengths (510 or 600 nm). This reaction could be applied for multinuclear complexes synthesized using various coupling reactions, i.e., the Mizoroki–Heck reactions, olefin metathesis, and Sonogashira coupling. This should be the first example of the photochemical hydrogenation reaction utilizing the photosensitizing abilities of the starting materials. It should be noted that the reaction proceeded even when the unsaturated linkers possessed bulky metal-complex units as substituents, which suppress thermal hydrogenation reactions using Pd/C or Crabtree catalysts. The hydrogenation reaction was also achieved with an electrochemical process. In the presence of a proton source and a base, the unsaturated linkers were hydrogenated during the bulk electrolysis. This result indicates that the hydrogenation reaction should be able to be achieved even when the multinuclear complexes do not have photosensitizing abilities. The photocatalytic abilities for CO_2 reduction of the multinuclear complexes synthesized using coupling reactions could be improved by the hydrogenation reaction. The new hetero-trinuclear photocatalysts **Os-Re-Ru** synthesized using the combination of the hydrogenation and stepwise Mizoroki–Heck reactions showed higher TON_{CO} than those of **Ru-Re** and **Os-Re**.

4.4 Experimental

General experimental procedures and synthetic procedures have been described in detail in the following journal articles published previously:

Y. Yamazaki, A. Umemoto, O. Ishitani, "Photochemical Hydrogenation of π-Conjugated Bridging Ligands on Photofunctional Multinuclear Complexes", *Inorg. Chem.*, **2016**, *55*, 11,110–11,124.

Y. Yamazaki, O. Ishitani, "Synthesis of Os(II)–Re(I)–Ru(II) hetero-trinuclear complexes and their photophysical properties and photocatalytic abilities", *Chem. Sci.*, **2018**, *9*, 1031–1041.

4.4.1 Electrochemical Reactions

The cyclic voltammograms of **Ru = Re** (0.01 mM) are measured in an MeCN solution or an MeCN/pyridine/CF_3COOH mixed solution (3:1:0.1 v/v/v) containing Et_4NBF_4 (0.1 M) as a supporting electrolyte using a HOKUTO DENKO HZ-7000 electrochemical analyzer with glassy-carbon disk working electrode (3 mm diameter), a Ag/$AgNO_3$ (0.01 M) reference electrode and a Pt counter electrode. The balk electrolysis of **Ru = Re** (0.01 mM) was conducted in an MeCN/pyridine/CF_3COOH mixed solution (3:1:0.1 v/v/v) containing Et_4NBF_4 (0.1 M) as a supporting electrolyte using a HOKUTO DENKO HZ-7000 electrochemical analyzer with mesh-texture glassy-carbon working electrode, a Ag/$AgNO_3$ (0.01 M) reference electrode, and a mesh-texture Pt counter electrode.

Acknowledgements Figures 4.6, 4.8, 4.9 and Table 4.1 were reprinted with permission from the following references:

Y. Yamazaki, A. Umemoto, O. Ishitani, "Photochemical Hydrogenation of π-Conjugated Bridging Ligands on Photofunctional Multinuclear Complexes", *Inorg. Chem.*, **2016**, *55*, 11,110–11,124.

Copyright 2016 American Chemical Society.

Figures 4.14 and 4.16 were reproduced from the following reference by permission of The Royal Society of Chemistry:

Y. Yamazaki, O. Ishitani, "Synthesis of Os(II)–Re(I)–Ru(II) hetero-trinuclear complexes and their photophysical properties and photocatalytic abilities", *Chem. Sci.*, **2018**, *9*, 1031–1041.

References

1. Connors PJ, Tzalis D, Dunnick AL, Tor Y (1998) Coordination compounds as building blocks: single-step synthesis of heteronuclear multimetallic complexes containing Ru^{II} and Os^{II}. Inorg Chem 37:1121
2. Tzalis D, Tor Y (1996) Coordination compounds as building blocks: single-step synthesis of multi-ruthenium(II) complexes. Chem Commun 1043
3. Haak RM, Martinez Belmonte M, Escudero-Adan EC, Benet-Buchholz J, Kleij AW (2010) Olefin metathesis as a tool for multinuclear Co(iii)salen catalyst construction: access to cooperative catalysts. Dalton Trans 39:593
4. Haak RM, Castilla AM, Martinez Belmonte M, Escudero-Adan EC, Benet-Buchholz J, Kleij AW (2011) Access to multinuclear salen complexes using olefin metathesis. Dalton Trans 40:3352
5. Arm KJ, Williams JAG (2006) A cross-coupling strategy for the synthesis of dimetallic assemblies containing mixed bipyridine-terpyridine bridging ligands: luminescence and energy transfer properties. Dalton Trans 2172
6. Arm KJ, Williams JAG (2005) Boronic acid-substituted metal complexes: versatile building blocks for the synthesis of multimetallic assemblies. Chem Commun 230
7. Whittle VL, Williams JAG (2009) Cyclometallated, bis-terdentate iridium complexes as linearly expandable cores for the construction of multimetallic assemblies. Dalton Trans 3929
8. Bian Z-Y, Chi S-M, Li L, Fu W (2010) Conjugation effect of the bridging ligand on the CO_2 reduction properties in difunctional photocatalysts. Dalton Trans 39:7884
9. Bian Z-Y, Wang H, Fu W-F, Li L, Ding A-Z (2012) Two bifunctional Ru^{II}/Re^{I} photocatalysts for CO_2 reduction: a spectroscopic, photocatalytic, and computational study. Polyhedron 32:78
10. Pac C, Ihama M, Yasuda M, Miyauchi Y, Sakurai H (1981) $Ru(bpy)_3{}^{2+}$-mediated photoreduction of olefins with 1-Benzyl-1,4-dihydronicotinamide: a mechanistic probe for electron-transfer reactions of NAD(P)H-Model compounds. J Am Chem Soc 103:6495
11. Tolman CA (1977) Steric effects of phosphorus ligands in organometallic chemistry and homogeneous catalysis. Chem Rev 77:313
12. van Houten J, Watts RJ (1978) Photochemistry of tris(2,2'-bipyridyl)ruthenium(II) in aqueous solutions. Inorg Chem 17:3381
13. Durham B, Caspar JV, Nagle JK, Meyer TJ (1982) Photochemistry of $Ru(bpy)_3{}^{2+}$ ion. J Am Chem Soc 104:4803
14. Kalyanasundaram K (1982) Photophysics, photochemistry and solar energy conversion with tris(bipyridyl)ruthenium(II) and its analogues. Coord Chem Rev 46:159
15. Tamaki Y, Koike K, Morimoto T, Ishitani O (2013) Substantial improvement in the efficiency and durability of a photocatalyst for carbon dioxide reduction using a benzoimidazole derivative as an electron donor. J Catal 304:22
16. Tamaki Y, Morimoto T, Koike K, Ishitani O (2012) Photocatalytic CO_2 reduction with high turnover frequency and selectivity of formic acid formation using Ru(II) multinuclear complexes. Proc Nat Acad Sci USA 109:15673
17. Tamaki Y, Watanabe K, Koike K, Inoue H, Morimoto T, Ishitani O (2012) Development of highly efficient supramolecular CO_2 reduction photocatalysts with high turnover frequency and durability. Faraday Discuss 155:115
18. Tamaki Y, Koike K, Morimoto T, Yamazaki Y, Ishitani O (2013) Red-light-driven photocatalytic reduction of CO_2 using Os(II)–Re(I) supramolecular complexes. Inorg Chem 52:11902

Chapter 5
Synthesis of Novel Multinuclear Complexes for Light-Harvesting Systems Using the Coupling Reactions with Ring-Shaped Multinuclear Re(I) Complexes as Building Blocks

Abstract As described in Chap. 3, reaction conditions for the selective progress of the Mizoroki–Heck reaction and the homo-coupling reaction were successfully established. The coupling reactions are potentially applied for connecting with not only mononuclear complexes but also multinuclear complexes. In this chapter, I chose ring-shaped Re(I) multinuclear complexes, which show suitable photophysical and electrochemical properties as photosensitizers, as building blocks for the coupling reactions. Re(I)-rings having a bromo group were synthesized using a new decarbonylation method with trimethylamine *N*-oxide and then linked with each other by the homo-coupling reaction or with a Ru(II) tris-diimine complex by the Mizoroki–Heck reaction. Because of the suitable photophysical properties of the Re(I)-ring units, the obtained new class of multinuclear complexes functioned as efficient light-harvesting systems.

Keywords Photofunctional multinuclear complexes · Coupling reactions · Intramolecular energy transfer · Light-harvesting functions

5.1 Introduction

In the previous chapters, I reported that mononuclear or dinuclear complexes possessing functional groups could be connected with other metal complexes by using various C–C coupling reactions. A combination of the coupling reactions enabled to synthesize novel multinuclear complexes to enlarge a variety of multinuclear complexes. However, the building block approach using coupling reactions has not yet been combined with the ordinary synthetic methods for multinuclear complexes, i.e., stepwise coordination of precursor complexes to pre-connected multidentate ligands as bridging ligands. Basically, multinuclear complexes linked via bridging ligands are difficult to be connected with other metal complexes because of the low reactivity for the ligand substitution reactions. If some functional groups can be introduced into the multinuclear complexes synthesized using the ordinary synthetic methods, such multinuclear complexes will be able to be connected with not only other mononuclear complexes but also other multinuclear complexes using coupling reactions in a single step by one-pot synthesis. By coupling the multinuclear

Y. Yamazaki, *Development of Synthetic Methods for Novel Photofunctional Multinuclear Complexes*, Springer Theses, https://doi.org/10.1007/978-981-16-7148-7_5

complexes with other metal complexes, useful photophysical properties and photofunctions of the multinuclear complexes will be enhanced and/or fused with other photofunctions to construct novel multifarious multinuclear complexes. Therefore, both introduction methods of functional groups into multinuclear complexes and integration methods for combing multinuclear complexes with other metal complexes should be developed and investigated.

Ring-shaped multinuclear Re(I) complexes should be one of the most interesting multinuclear complexes, which attract interest in recent years because of their unique photophysical properties [1], efficient photosensitizing abilities [2], and abilities for multiple electron accumulation [3]. Such properties of Re(I) rings are desirable for utilizing as units of multifarious multinuclear complexes with high efficiencies.

In this chapter, I will focus on the Re(I) rings as a typical example of multinuclear complexes synthesized using the ordinary synthetic methods and will report synthetic methods for both introducing functional groups into Re(I) rings and linking them with other metal complexes. By modifying the decarbonylation step, which is required to accept coordination of bidentate phosphine ligands for cyclization, a Re(I) ring possessing a bromo group could be successfully synthesized. This Re(I) ring could be connected with a Ru(II) tris-diimine complex and another Re(I) ring using the Mizoroki–Heck reaction and the homo-coupling, respectively. These coupling reactions proceeded selectively under the same reaction condition described in Chap. 3. These results clearly indicated that the building block approach could be applied even for the connection of multinuclear complexes with other metal complexes. The obtained multinuclear complexes possessing Re(I) ring moieties showed efficient intramolecular energy transfer. The energy transfer took place almost quantitatively, which led to achieving light-harvesting systems.

5.2 Results and Discussion

5.2.1 Decarbonylation Reactions Using Me_3NO

In the synthesis of Re(I) rings, both terminal CO ligands of the corresponding linear-shaped multinuclear complexes are removed by a photochemical ligand substitution reaction before the cyclization steps (Eq. 5.1) [1–3]. This photochemical reaction often requires a long irradiation time, which causes various decomposition reactions. The slow reaction rate is likely because the photochemical ligand substitution reaction is suppressed by intramolecular energy transfer from tricarbonyl complex units to bis-carbonyl complex units. The decomposition reactions caused by long-time light irradiation lowered the isolation yields for the products, especially when the bidentate phosphine ligands (P–P) have short carbon chains, such as ethylene (1,2-bis(diphenylphosphino)ethylene: dppe) and propylene (1,3-bis(diphenylphosphino)propylene: dpppr). Besides, photochemical ligand substitution hardly occurred when Re complexes have electron-withdrawing groups, e.g., bromo groups, because of the suppression of inter-conversion from the triplet metal-to-ligand charge transfer (^{3}MLCT) excited state to the triplet ligand field (^{3}LF) excited state [4]. Therefore, in order to introduce bromo groups into Re(I) rings, it is required to develop a new decarbonylation method instead of the photochemical ligand substitution.

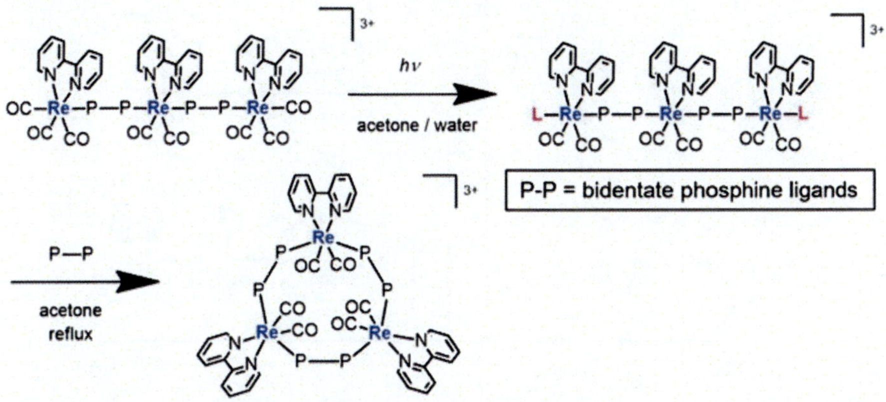

(5.1)

Some decarbonylation reactions of various carbonyl complexes using a thermal reaction with Me_3NO have been reported [5–8]. In this reaction, CO ligands will be oxidized to eliminate the ligands as CO_2, followed by coordination of ambient solvent molecules or Me_3N. This thermal reaction could also be applied for removing the CO ligands of Re(I) mononuclear tri-carbonyl complexes (Eq. 5.2). The tri-carbonyl complexes were quantitatively converted to the bis-carbonyl complexes within 1-h refluxing regardless of substituents on phosphine ligands, e.g., PPh_3 and PEt_3. By using this reaction, various tri-carbonyl Re(I) complexes could be selectively converted to bis-carbonyl complexes even when complexes having electron-withdrawing groups, e.g., trifluoromethyl groups or bromo groups.

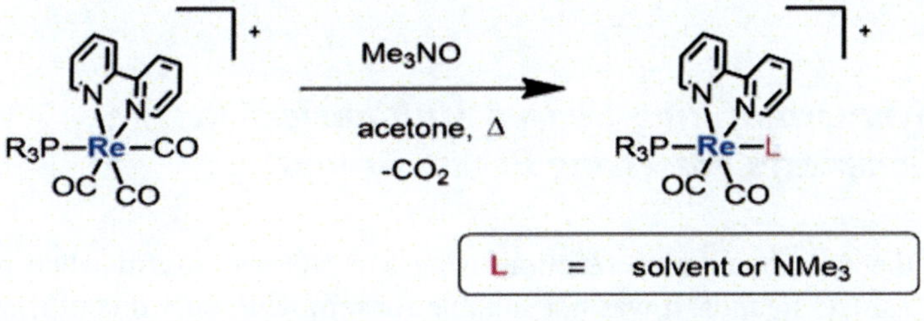

(5.2)

This reaction was useful even for decarbonylation of both terminal CO ligands on the linear-shaped multinuclear Re(I) complexes (Fig. 5.1). As a typical example, a linear-shaped trinuclear Re(I) complexes ([{Re(bpy)$(CO)_3(\mu_2$-dppa)}$_2$Re(bpy)$(CO)_2]^{3+}$, dppa = bis(diphenylphosphino)acetylene) and Me_3NO were dissolved in acetone and the solution was heated. As soon as the addition of

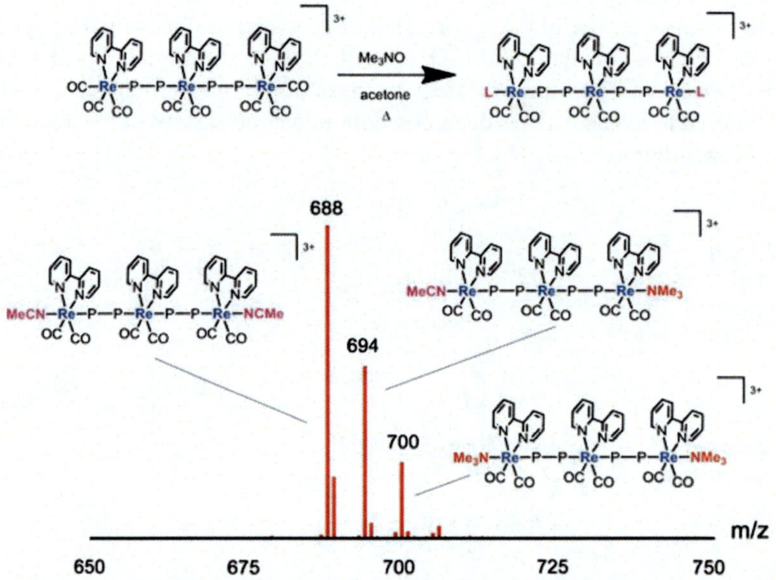

Fig. 5.1 ESI-mass spectrum measured after the reaction between Me_3NO and the linear-shaped trinuclear Re(I) complex with dppa. The mobile phase was MeCN

Me_3NO, the color of the solution was changed from yellow to orange. The ESI-mass spectrum recorded after 10 min of heating revealed that all of the Re(I) trinuclear complexes were converted to the products, of which two terminal CO ligands were substituted with ambient solvent molecules or Me_3N. The decarbonylation occurred only on tri-carbonyl complex units in both edges of the trinuclear complex without any side reaction. The reaction proceeded under relatively mild conditions, i.e., even at room temperature. These results clearly show that the decarbonylation reaction with Me_3NO is a useful synthetic method not only for decreasing the reaction time but also for suppressing side reactions.

5.2.2 *Synthesis of Ring-Shaped Multinuclear Re(I) Complexes Possessing Bromo Groups*

Although the decarbonylation reaction using Me_3NO was useful when removing *both* terminal CO ligands, it was not suitable for removing only one of the terminal CO ligands. Therefore, I used a combination of the photochemical ligand substitution and the decarbonylation with Me_3NO to obtain trinuclear complexes with a bromo group selectively. Scheme 5.1 shows the synthetic scheme for a hetero-trinuclear Re(I) ring having a bromo group. Firstly, one of the terminal CO ligands in a dinuclear complex was removed by the photochemical ligand substitution within

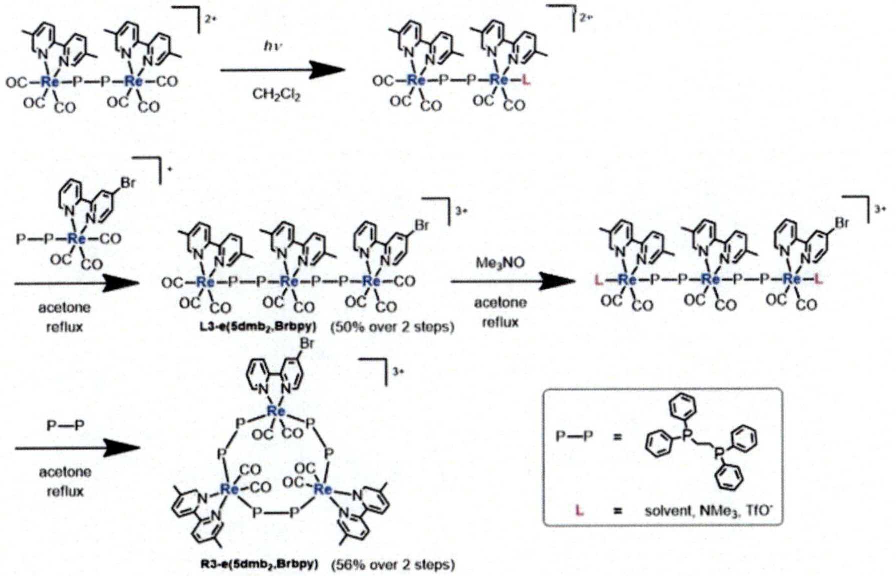

Scheme 5.1 Synthetic scheme of the ring-shaped hetero-trinuclear Re(I) complexes with a bromo group

30 min of irradiation. Next, the decarbonylated dinuclear complex was reacted with a mononuclear complex having a bromo group and a bidentate phosphine ligand, of which one side maintains the coordination ability. The terminal CO ligands in the obtained linear-shaped trinuclear complex were then decarbonylated by Me_3NO and the subsequent cyclization with dppe gave the targeted Re(I) ring with a bromo group with a 56% yield.

These results clearly indicate that the synthetic strategy mentioned above can be applied for the synthesis of various Re(I) rings with electron-withdrawing groups.

5.2.3 Coupling Reactions Using Re(I) Rings Possessing Bromo Groups as Starting Materials

As described in the preceding section, the ring-shaped multinuclear Re(I) complex having a bromo group (**R3-e(5dmb$_2$,Brbpy)**, 5dmb = 5,5′-dimethyl-2,2′-bipyridine, Brbpy = 4-bromo-2,2′-bipyridine) was successfully synthesized with a relatively high isolated yield by using two kinds of decarbonylation reactions. Therefore, new types of multinuclear complexes possessing Re(I) ring moieties were expected to be synthesized with **R3-e(5dmb$_2$,Brbpy)** as a starting material using various coupling reactions. As typical examples, I tried the homo-coupling and the Mizoroki–Heck reaction using **R3-e(5dmb$_2$,Brbpy)** as building blocks for the building block approach described in the previous chapters (Eqs. 5.3 and 5.4).

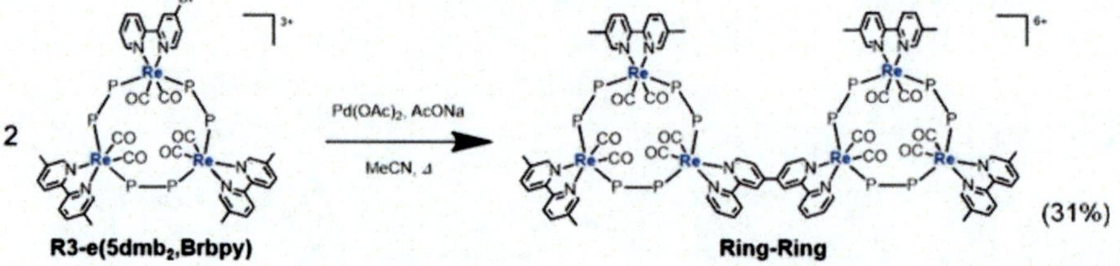

(5.3)

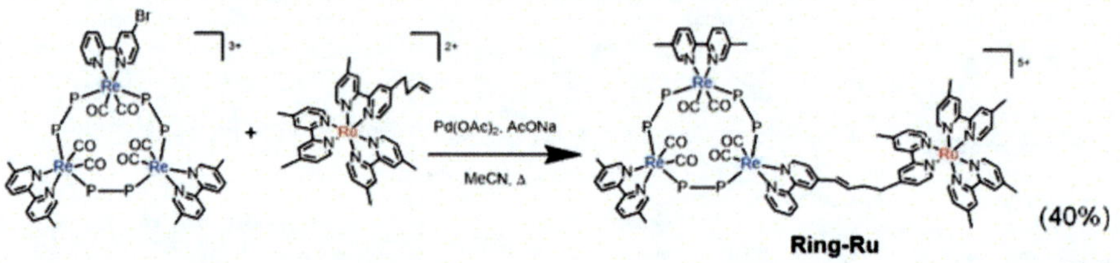

(5.4)

In the case of the homo-coupling reaction, the reaction rate was so slow that precipitation of Pd black as a competing reaction was obviously observed, and the reaction was stopped within one day. This is probably because the reaction rate of the homo-coupling reaction strongly depends on the steric hindrance of substrates, and bulky **R3-e(5dmb$_2$,Brbpy)** lowered the reaction rate compared to the production of Pd black. In order to suppress the acceleration of production of Pd black, the precipitate was filtered off through a membrane filter (Millex LG 0.20 μm filter), and an additional MeCN solution containing $Pd(OAc)_2$ and AcONa was then added into the filtrate. This process was repeated seven times. Figure 5.2 shows the chart of size

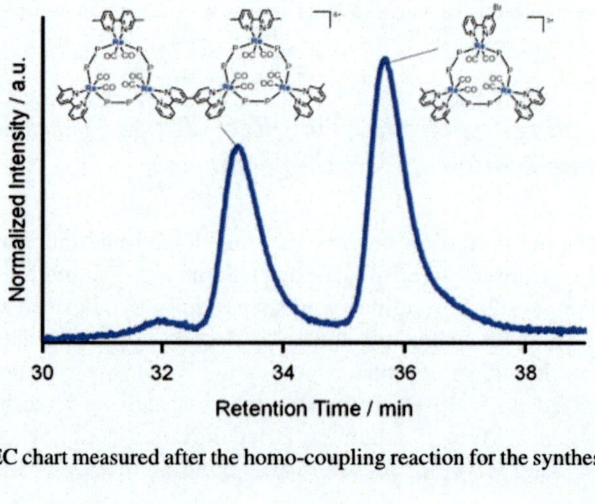

Fig. 5.2 The SEC chart measured after the homo-coupling reaction for the synthesis of **Ring-Ring**

exclusion chromatography (SEC) measured after the reaction. In the chromatogram, a new peak, which was attributable to the target product, was observed at a faster retention time (RT = 33 min) than that of **R3-e(5dmb$_2$,Brbpy)** (RT = 36 min). The chromatogram also revealed that the reaction proceeded with high selectivity even though the conversion of the reaction was not so high. The conversion yield was determined to be 51% by SEC. The targeted dimer of the Re(I) ring **(Ring-Ring)** was isolated with a 31% yield using preparative SEC.

The Mizoroki–Heck reaction between **R3-e(5dmb$_2$,Brbpy)** and two equivalents of a Ru complex having a butenyl group proceeded in a MeCN solution containing $Pd(OAc)_2$ and AcONa under reflux. In this reaction, an excess amount of the Ru complex was used to suppress the homo-coupling reaction. The reaction rate was higher compared to that of the homo-coupling reaction, probably because of relatively low steric hindrance around the vinyl group of the Ru complex. As described above, Pd black was precipitated after 1 day. The precipitate was removed by a membrane filter and the solvent was evaporated before the addition of MeCN solution containing $Pd(OAc)_2$ and AcONa. This solution was again refluxed for 2 days. These processes were repeated twice. Figure 5.3 shows the SEC charts recorded after the reaction. The targeted tetranuclear complex (**Ring-Ru**) and the two starting complexes were mainly detected in the chromatogram. This result suggested that the reaction selectivity was high. The conversion yield of **Ring-Ru** was estimated to be 64% from the chromatogram. The target product was isolated with a 40% yield using preparative SEC. From these results, It can be concluded that the Re(I) rings with a bromo group can be used as building blocks for the building block approach.

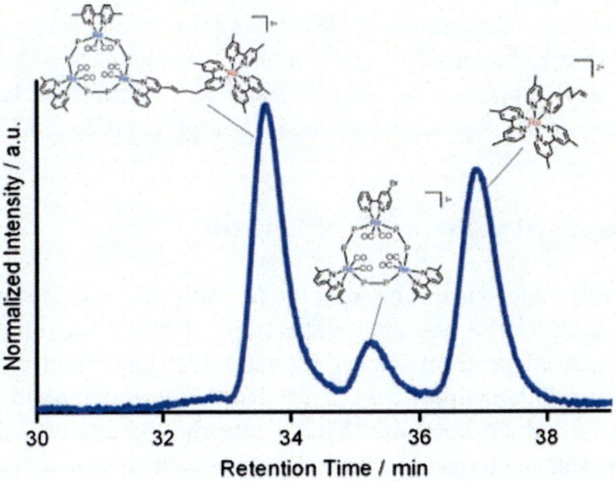

Fig. 5.3 The SEC chart measured after the Mizoroki–Heck reaction for the synthesis of **Ring-Ru**

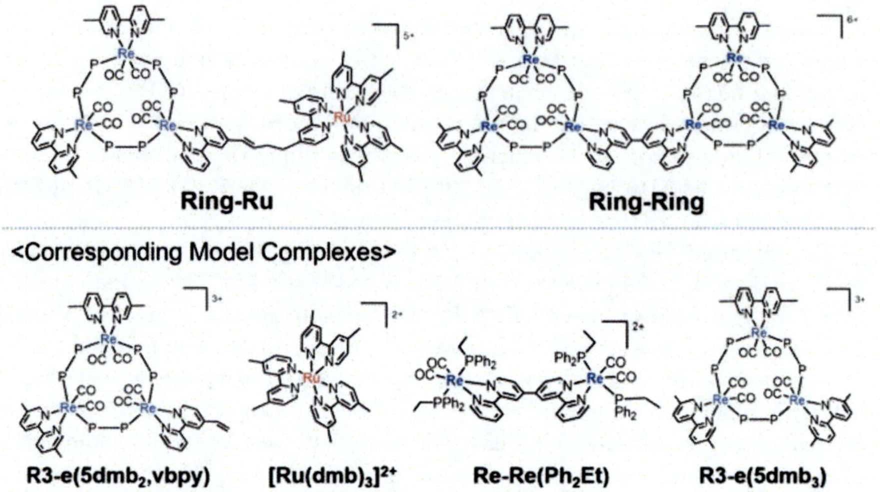

Chart 5.1 The structures of **Ring-Ru**, **Ring-Ring**, and their corresponding model complexes

5.2.4 Photophysical Properties and Light-Harvesting Abilities of Ring-Ru and Ring-Ring

The obtained **Ring-Ru** and **Ring-Ring** have two kinds of photofunctional complex moieties, i.e., the Re(I) ring unit and the Ru unit in **Ring-Ru**, and two kinds of Re(I) units with different diimine ligands in **Ring-Ring**, respectively. Therefore, these multinuclear complexes might show efficient intramolecular energy transfer between the different units to function as light-harvesting systems. Therefore, I investigated the photophysical properties of **Ring-Ru** and **Ring-Ring** in detail by comparing them with those of the corresponding model complexes (Chart 5.1).

5.2.4.1 Photophysical Properties of Ring-Ru

Figure 5.4 shows the absorption spectra of **Ring-Ru** and its model complexes measured in MeCN. The absorption spectrum of **Ring-Ru** (blue line) is similar to a 1:1 summation spectrum of model complexes (black dotted line). This result indicates that the interaction between the Re(I)-ring moiety and the Ru complex moiety is not strong. Of note, the ^{1}MLCT absorption band of the Ru(II) unit (red line) was observed at a longer wavelength region compared with that of the ring unit (green line).

The emission spectra of **Ring-Ru** and the model complexes are shown in Fig. 5.5a. When measured with $\lambda_{ex} = 380$ nm, the emission intensity at the shorter wavelength region became larger compared with that measured with $\lambda_{ex} = 460$ nm (Fig. 5.5b). Both the Re(I)-ring and the Ru units absorb the light at $\lambda_{ex} = 380$ nm, whereas the Ru unit is excited selectively with the light at $\lambda_{ex} = 460$ nm. This result indicates

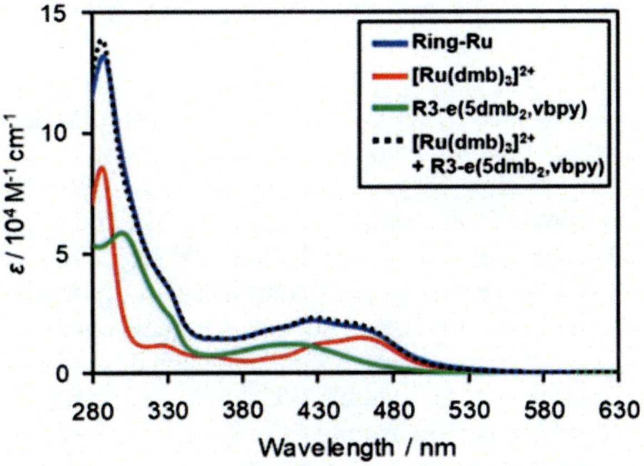

Fig. 5.4 UV–vis absorption spectra of **Ring-Ru** and the corresponding model complexes in MeCN (blue: **Ring-Ru**, red: $[Ru(dmb)_3]^{2+}$, green: **R3-e(5dmb$_2$, vbpy)**, black dotted line: the summation spectrum of $[Ru(dmb)_3]^{2+}$ and **R3-e(5dmb$_2$, vbpy)**)

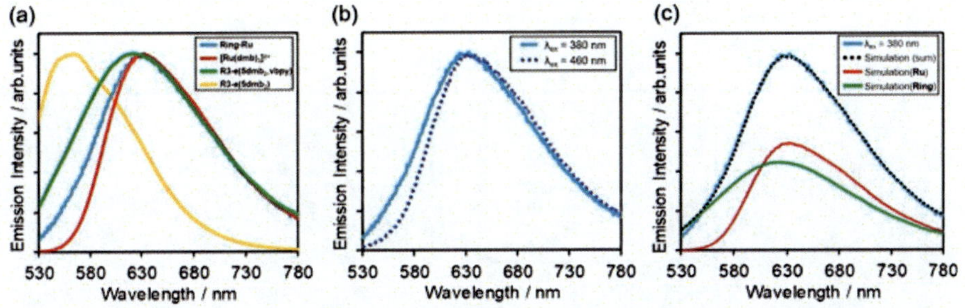

Fig. 5.5 **a** Normalized emission spectra of **Ring-Ru** in degassed MeCN (the excitation wavelength is $\lambda_{ex} = 380$ nm (light blue). The emission spectra of $[Ru(dmb)_3]^{2+}$ (red), **R3-e(5dmb$_2$, vbpy)** (green), and **R3-e(5dmb$_3$)** (orange) are also illustrated. **b** Normalized emission spectra of **Ring-Ru** with different exciatation wavelength (the excitation wavelength is $\lambda_{ex} = 380$ (light blue) or 460 nm (blue dotted line). **c** Emission spectra of **Ring-Ru** ($\lambda_{ex} = 380$ nm, light blue) and simulation of spectral deconvolution into individual spectra of $[Ru(dmb)_3]^{2+}$ (red), **R3-e(5dmb$_2$, vbpy)** (green) and their sum (black dotted line)

that the emission spectrum recorded with $\lambda_{ex} = 380$ nm contains emission from both the Re(I)-ring and the Ru units. The emission spectrum recorded with $\lambda_{ex} =$ 380 nm could be well-simulated with the emission spectra of **R3-e(5dmb$_2$,vbpy)** and $[Ru(dmb)_3]^{2+}$ (Fig. 5.5c). The emission quantum yields (Φ_{em}s) also depended on the excitation wavelength. The emission quantum yield determined with $\lambda_{ex} =$ 460 nm ($\Phi_{em} = 8.6\%$) was similar to that of $[Ru(dmb)_3]^{2+}$ ($\Phi_{em} = 8.9\%$), whereas the emission quantum yield determined with $\lambda_{ex} = 380$ nm ($\Phi_{em} = 6.0\%$) became the value between that of $[Ru(dmb)_3]^{2+}$ and **R3-e(5dmb$_2$,vbpy)** ($\Phi_{em} = 4.2\%$).

5.2.4.2 Intramolecular Energy Transfer and Light-Harvesting Abilities of Ring-Ru

Before investigating the energy transfer in **Ring-Ru**, the photophysical behaviors of **R3-e(5dmb$_2$,vbpy)** was carefully checked because the ring moiety consists of two different Re units, i.e., the Re units with 5dmb and that with vbpy (vbpy = 4-vinyl-2,2′-bipyridine), and intramolecular energy transfer might occur among the units. Figure 5.6a shows the transient emission spectra of **R3-e(5dmb$_2$,vbpy)**. The maximum emission wavelength gradually shifted from $\lambda_{max} = 605$ nm (5 ns) to $\lambda_{max} = 635$ nm (1000 ns). It should be noted the maximum emission wavelength of **R3-e(5dmb$_3$)**, which is a model compound of the Re unit with 5dmb, was much shorter than the λ_{max} at 1000 ns (Figs. 5.5a and 5.6b); therefore, the emission at 1000 ns should be derived mainly from the Re unit with vbpy. The emission decay at $\lambda_{det} = 550$ nm was much faster than that at $\lambda_{max} = 650$ nm (Fig. 5.6c). The emission decay curves could be fitted with a double exponential function; the obtained lifetimes were

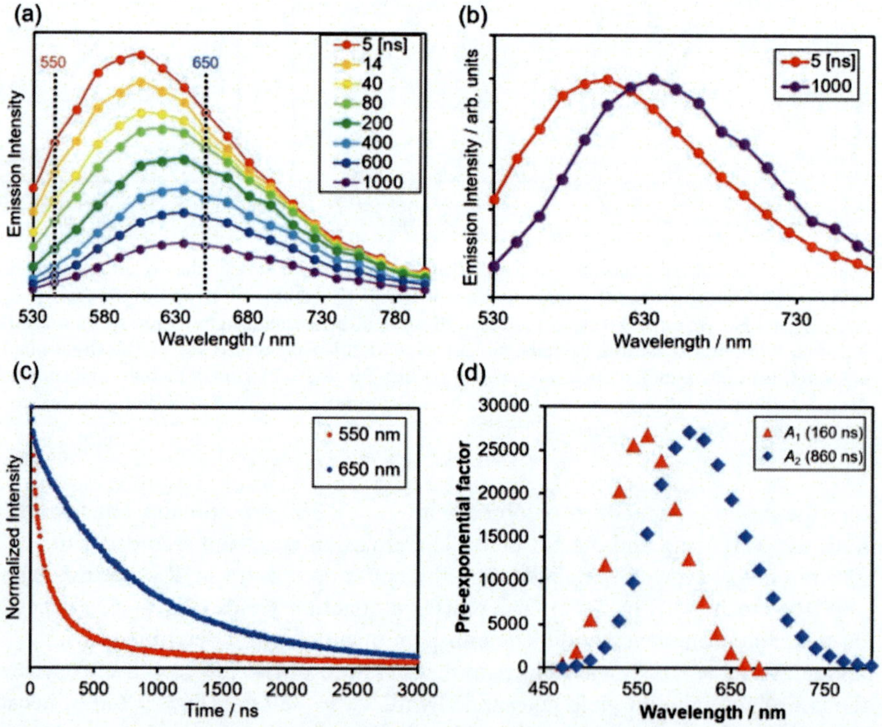

Fig. 5.6 **a** Transient emission spectra of **R3-e(5dmb$_2$, vbpy)** measured in degassed MeCN with $\lambda_{ex} = 444$ nm. **b** Normalized emission spectra recorded at 5 (red) and 1000 ns (purple) after the laser excitation. **c** Emission decay curves of **R3-e(5dmb$_2$, vbpy)** recorded at 550 (red) and 650 nm (blue). **d** Wavelength dependence of the pre-exponential factors of τ_1 (A_1, red triangles) and τ_2 (A_2, blue squares)

$\tau_1 = 160$ ns and $\tau_2 = 860$ ns. The wavelength dependence of the pre-exponential factors of τ_1 was similar to the emission spectrum of **R3-e(5dmb$_3$)** (Fig. 5.6d); therefore, τ_1 and τ_2 could be attributed to the emission lifetimes associated with the Re units with 5dmb and that with vbpy, respectively. These results indicate that the intramolecular energy transfer proceeded from the Re unit with 5dmb to that with vbpy. The rate constant for the intramolecular energy transfer was determined to be $k_{ET1} = 6.1 \times 10^6$ s^{-1} using Eq. 5.5 with τ_1 and the emission lifetime of **R3-e(5dmb$_3$)** ($\tau_{(model)} = 5.4$ µs). The efficiency of the intramolecular energy transfer was estimated to be $\eta_{ET1} = 97\%$ based on Eq. 5.6 using k_{ET1} and $\tau_{(model)}$. This value shows that the intramolecular energy transfer in **R3-e(5dmb$_2$,vbpy)** proceeds efficiently.

$$\begin{aligned} k_{ET1} &= 1/\tau_{1(R3-e(5dmb2,vbpy))} - 1/\tau_{(model)} \\ &= 1/(160\ \mathrm{ns}) - 1/(5.4\mu\mathrm{s}) \\ &= 6.1 \times 10^6\left[\mathrm{s}^{-1}\right] \end{aligned} \tag{5.5}$$

$$\begin{aligned} \eta_{ET1} &= k_{ET1}\big/\left(1/\tau_{1(R3-e(5dmb2,vbpy))}\right) \\ &= \left(6.1 \times 10^6\mathrm{s}^{-1}\right)\big/\{1/(160\ \mathrm{ns})\} \\ &= 97\% \end{aligned} \tag{5.6}$$

Figure 5.7a shows the transient emission spectra of **Ring-Ru**. The emission intensity in the shorter wavelength region (< 600 nm) decreased within 50 ns. The rate of this emission decay was much faster compared with the transient emission spectra of **R3-e(5dmb$_2$,vbpy)**. This behavior should be derived from the additional energy transfer from the Re units with 5dmb, i.e., the excitation energy of the Re units with 5dmb was transferred not only to the Re unit with vbpy but also to the Ru unit. Besides, from 50 to 400 ns, another slower and smaller change in the spectrum shape was observed, i.e., a decrease in the shorter wavelength region (< 600 nm) and an increase around 630 nm (Fig. 5.7b). This spectral change was attributed to a slow energy transfer process from the Re unit with vbpy to the Ru unit because the excitation energy of the Ru unit ($\lambda_{em} = 633$ nm) is smaller than that of the Re unit with vbpy ($\lambda_{em} = 622$ nm). After 400 ns, the spectrum shape did not change, and the emission intensity was monotonically decreased. Interestingly, the spectrum recorded after 800 ns contained both the emission from **R3-e(5dmb$_2$,vbpy)** and that from the Ru unit (Fig. 5.8). These results clearly indicate that the energy transfer between the Re unit with vbpy and the Ru unit proceeded reversibly, and the energy transfer processes should arrive at an equilibrium state at 400 ns after the laser irradiation. The emission decay curves of **Ring-Ru** are shown in Fig. 5.7c. The decay curves recorded at $\lambda_{em} = 550$ and 650 nm as typical examples could be roughly fitted using a triple exponential function with $\tau_1 = 23$ ns, $\tau_2 = 140$ ns, and $\tau_3 = 980$ ns. Of note, the pre-exponential factors of τ_2 (A_2) at the longer wavelength region (> 600 nm) were negative because of the slower energy transfer from the Re unit with vbpy to the Ru unit between 50 and 400 ns (Fig. 5.7d).

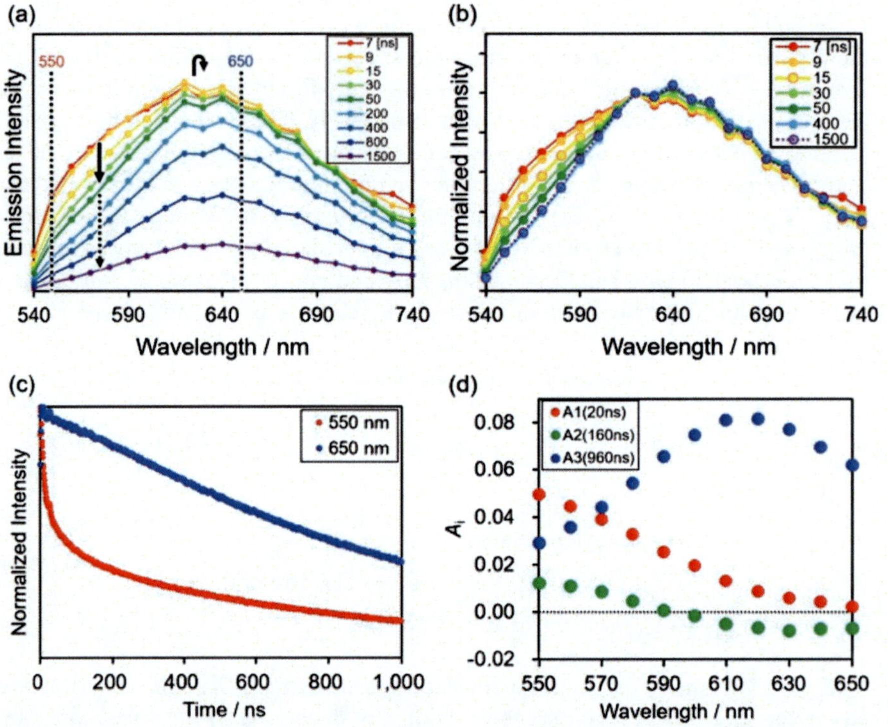

Fig. 5.7 **a** Transient emission spectra of **Ring-Ru** measured in degassed MeCN with $\lambda_{ex} = 337$ nm. **b** Transient emission spectra of **Ring-Ru** normalized at 620 nm. **c** Emission decay curves of **R3-e(5dmb$_2$, vbpy)** recorded at 550 (red) and 650 nm (blue). **d** Wavelength dependence of the pre-exponential factors of τ_1 (A_1, red circles), τ_2 (A_2, green circles), and τ_3 (A_3, blue circles)

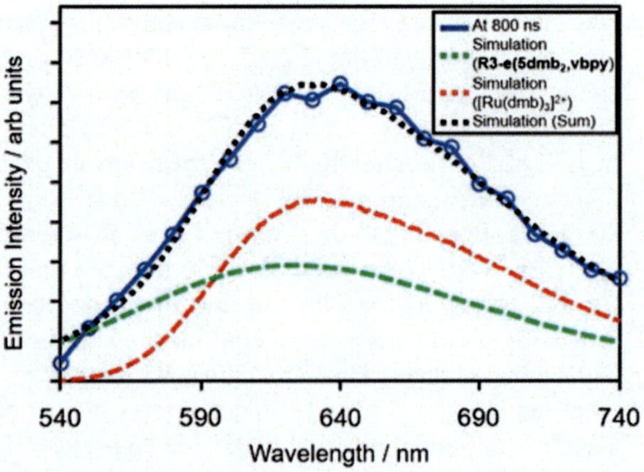

Fig. 5.8 Transient emission spectra of **Ring-Ru** recorded at 800 ns (blue solid line) after the laser excitation and simulation of spectral deconvolution into individual spectra of $[Ru(dmb)_3]^{2+}$ (red broken line), **R3-e(5dmb$_2$, vbpy)** (green broken line) and their sum (black dotted line)

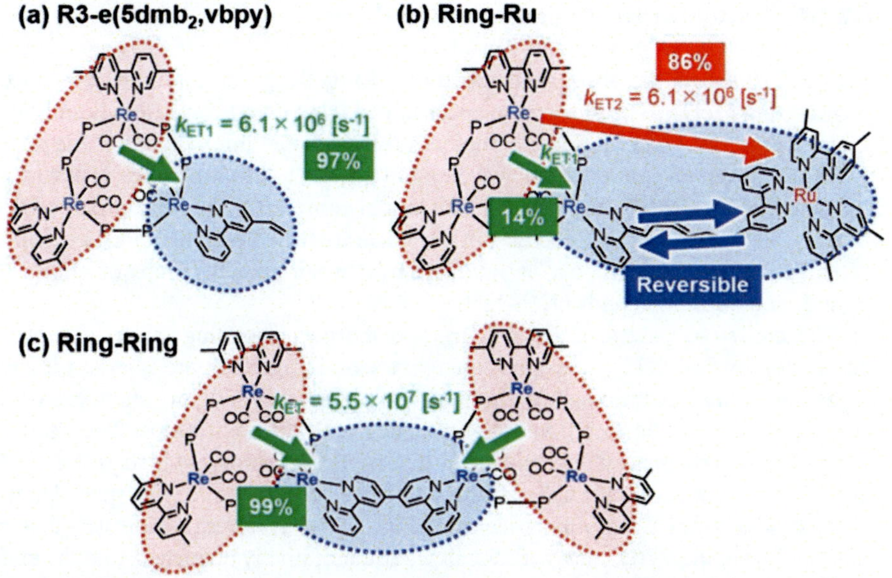

Fig. 5.9 Energy transfer processes and their rate constants in **a R3-e(5dmb,vbpy)**, **b Ring-Ru**, and **c Ring-Ring**

The rate constant of the energy transfer from the Re units with 5dmb to the Ru units (k_{ET2}) was roughly estimated to be 3.7×10^7 [s^{-1}] using Eq. 5.7 with τ_1, $\tau_{(model)}$, and k_{ET1}. According to Eqs. 5.8 and 5.9, the excitation energy of the Re unit with 5dmb was almost quantitatively transferred partially to the Re unit with vbpy and mostly to the Ru unit. These excitation energy transfer events are summarized in Fig. 5.9.

$$\begin{aligned} k_{ET2} &= 1/\tau_{1(\textbf{Ring-Ru})} - 1/\tau_{(model)} - k_{ET1} \\ &= 1/(23\ \text{ns}) - 1/(5.4\mu\text{s}) - 6.1 \times 10^6 \\ &= 3.7 \times 10^7\left[\text{s}^{-1}\right] \end{aligned} \tag{5.7}$$

$$\begin{aligned} \eta_{all} &= (k_{ET1} + k_{ET2})/(1/\tau_{1(\textbf{Ring-Ru})}) \\ &= \left(4.3 \times 10^7\text{s}^{-1}\right)/\{1/(23\ \text{ns})\} > 99\% \end{aligned} \tag{5.8}$$

$$\begin{aligned} \eta_{ET2} &= k_{ET2}/(1/\tau_{1(\textbf{Ring-Ru})}) \\ &= \left(6.1 \times 10^6\text{s}^{-1}\right)/\{1/(23\ \text{ns})\} \\ &= 86\% \end{aligned} \tag{5.9}$$

5.2.4.3 Photophysical Properties of Ring-Ring

Figure 5.10 shows the absorption spectra of **Ring-Ring** and its model complexes measured in MeCN. The absorption spectrum of **Ring-Ring** (blue line) is similar to a summation spectrum of model complexes (black dotted line), though some differences are observed in the longer wavelength region (> 420 nm). This result might indicate that **Re–Re(Ph$_2$Et)**, which is the model complex of the Re complex moiety having the bridging ligands, could not demonstrate the effect of cyclization, which often causes intramolecular π–π interaction between phenyl groups of phosphine ligands and diimine ligands [1, 9–11].

The emission spectra of **Ring-Ring** and the corresponding model complexes are shown in Fig. 5.11. The emission spectra of **Ring-Ring** are almost identical regardless of the excitation wavelength, i.e., 420, 450, and 480 nm, which can excite two kinds of different Re(I) complex moieties with different ratios. The shape of the spectra is completely different from that of **R3-e(5dmb$_3$)**, which is the model complex of the Re complex moiety with 5dmb. Besides, the excitation spectrum was quite similar to the absorption spectrum (Fig. 5.12). This result indicated that the emission from the Re(I) units with 5dmb was quantitatively quenched, and the emission of **Ring-Ring** was mostly derived from the central Re(I) units with the bridging ligands. This phenomenon was similar to **R3-e(5dmb$_2$,vbpy)**, which showed efficient intramolecular energy transfer.

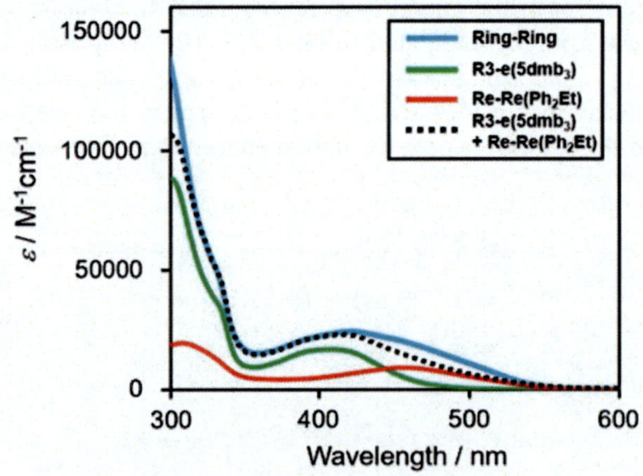

Fig. 5.10 UV–vis absorption spectra of **Ring-Ring** and the corresponding model complexes in MeCN (light blue: **Ring-Ring**, green: **R3-e(5dmb$_3$)**, red: **Re–Re(Ph$_2$Et)**, black dotted line: 1:4/3 summation spectrum of **Re–Re(Ph$_2$Et)** and **R3-e(5dmb$_3$)**

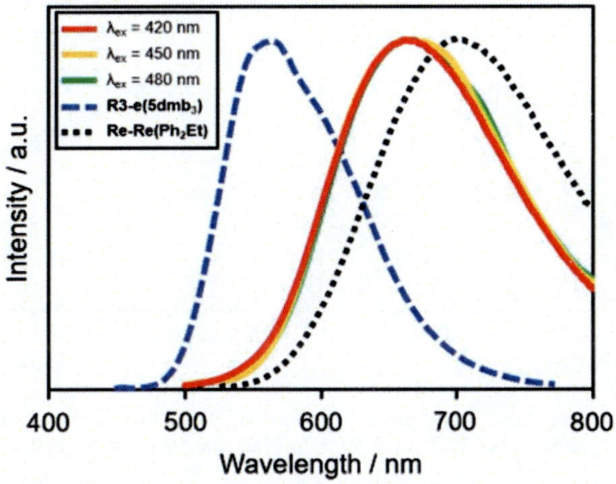

Fig. 5.11 Normalized emission spectra of **Ring-Ring** in degassed MeCN measured using various excitation wavelengths: $\lambda_{ex} = 420$ nm (red), $\lambda_{ex} = 450$ nm (orange), and $\lambda_{ex} = 480$ nm (green). The emission spectra of **R3-e(5dmb$_3$)** (blue broken line) and **Re–Re(Ph$_2$Et)** (black dotted line) are also illustrated

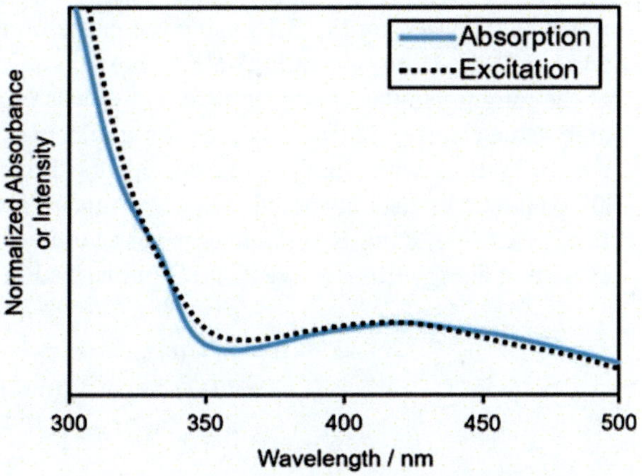

Fig. 5.12 Normalized absorption (light blue solid line) and the excitation spectrum of **Ring-Ring** (black dotted line). The detection wavelength was 630 nm

5.2.4.4 Intramolecular Energy Transfer and Light-Harvesting Abilities of Ring-Ring

The transient emission spectra of **Ring-Ring** are shown in Fig. 5.13a. The shape of the spectra was changed within 40 ns after excitation, and the intensity decreased gradually without a drastic change in the shape of the spectra. Especially in the shorter wavelength region, rapid decay of intensity was observed. Emission decay curves observed at 550 and 750 nm are illustrated in Fig. 5.13b. In the case of 550 nm, quite rapid decay was observed within several tens of ns, while a slight increase was observed at $\lambda_{ex} = 750$ nm. The emission decay curves could be fitted using the same

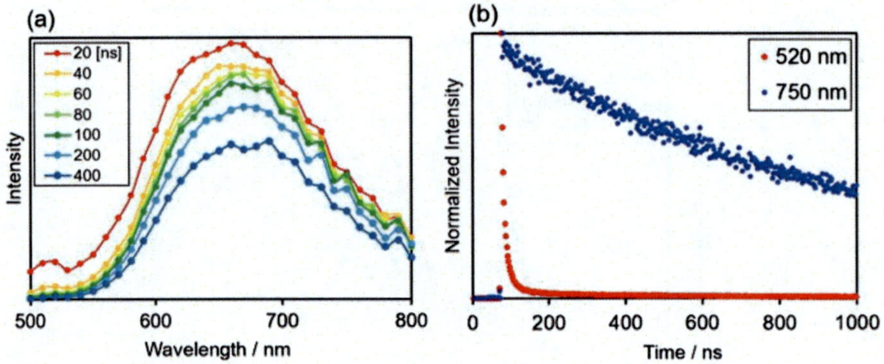

Fig. 5.13 **a** Transient emission spectra in degassed MeCN. **b** Emission decay curves of **Ring-Ring** recorded at 520 (red line) and 750 nm (blue line). The excitation wavelength was 444 nm

double exponential functions, and the lifetimes were $\tau_1 = 21$ ns and $\tau_2 = 993$ ns, respectively. The shorter lifetime (τ_1) was attributed to the lifetime of the Re(I) units with 5dmb, because the pre-exponential factors of τ_1 (A_1) became larger in the shorter wavelength region, and λ_{em} of **R3-e(5dmb$_3$)** is shorter than that of **Re–Re(Ph$_2$Et)**. The rate constant of the intramolecular energy transfer (k_{ET}) was estimated to be 5.5 × 10^7 [s^{-1}] based on Eq. 5.10 using the emission lifetimes of the Re(I) units with 5dmb in **Ring-Ring** (τ_1) and **R3-e(5dmb$_3$)** ($\tau_{(model)}$). By using the k_{ET}, the efficiency of the intramolecular energy transfer (η_{ET}) was estimated at 99% (Eq. 5.11). This high value clearly indicated that the intramolecular energy transfer proceeded from the Re(I) units with 5dmb to those with the bridging ligands quantitatively, and all excitation energy absorbed was accumulated into the center moieties of **Ring-Ring**; therefore, **Ring-Ring** functioned as an efficient light-harvesting system. These excitation energy transfer events are summarized in Fig. 5.9c.

$$\begin{aligned} k_{ET} &= 1/\tau_{1(\mathbf{Ring-Ring})} - 1/\tau_{(model)} \\ &= 1/(21\ \text{ns}) - 1/(5.4\ \mu\text{s}) \\ &= 5.5 \times 10^7 \left[\text{s}^{-1}\right] \end{aligned} \tag{5.10}$$

$$\begin{aligned} \eta_{ET} &= k_{ET}/(1/\tau_{(donor)}) \\ &= \left(5.5 \times 10^7 \text{s}^{-1}\right)/\{1/(21\ \text{ns})\} \\ &= 99\% \end{aligned} \tag{5.11}$$

5.3 Conclusions

In this chapter, I firstly reported that the decarbonylation reaction using Me_3NO was useful for converting Re(I) tri-carbonyl complexes to bis-carbonyl complexes. The reaction proceeded selectively and quantitatively and could be applied for not only mononuclear complexes but also linear-shaped multinuclear complexes. By using this thermal decarbonylation reaction, a new ring-shaped multinuclear Re(I) complex possessing a bromo group could be synthesized with a relatively high isolation yield. This Re(I) ring was successfully used as a starting material for the building block approach to connect with a Ru complex or a Re(I) ring using the coupling reactions, i.e., the Mizoroki–Heck reaction and homo-coupling. These results indicated that the ordinary synthetic methods for multinuclear complexes using multidentate bridging ligands could be combined with the building block approach to develop a new class of multifarious multinuclear complexes. The synthesized multinuclear complexes with Re(I) ring moieties, i.e., **Ring-Ru** and **Ring-Ring**, showed intramolecular energy transfer to function as efficient light-harvesting systems.

5.4 Experimental

General experimental procedures and synthetic procedures have been described in detail in the following journal articles published previously:

Y. Yamazaki, J. Rohacova, H. Ohtsu, M. Kawano, O. Ishitani, "Synthesis of Re(I) Rings Comprising Different Re(I) Units and Their Light-Harvesting Abilities", *Inorg. Chem.*, **2018**, *57*, 15,158–15,171.

Y. Yamazaki, O. Ishitani, "Synthesis of an Emissive Spectacle-Shaped Hexanuclear Rhenium(I) Complex", *Inorg. Chem.*, **2019**, *58*, 12,905–12,910.

Y. Yamazaki, J. Rohacova, K. Koike, O. Ishitani, "Synthesis and Light-Harvesting Functions of Ring-Shaped Re(I) Trinuclear Complexes Connected with an Emissive Ru(II) Complex", *JACS Au.*, **2021**, *1*, 294–307.

Acknowledgements Figures 5.10, 5.11, and 5.13 were reprinted with permission from the following references:

Y. Yamazaki, O. Ishitani, "Synthesis of an Emissive Spectacle-Shaped Hexanuclear Rhenium(I) Complex", *Inorg. Chem.*, **2019**, *58*, 12,905–12,910.

References

1. Rohacova J, Sekine A, Kawano T, Tamari S, Ishitani O (2015) Trinuclear and tetranuclear Re(I) rings connected with phenylene, vinylene, and ethynylene chains: synthesis, photophysics, and redox properties. Inorg Chem 54:8769–8777

2. Morimoto T, Nishiura C, Tanaka M, Rohacova J, Nakagawa Y, Funada Y, Koike K, Yamamoto Y, Shishido S, Kojima T, Saeki T, Ozeki T, Ishitani O (2013) Ring-shaped Re(I) multinuclear complexes with unique photofunctional properties. J Am Chem Soc 135:13266–13269
3. Asatani T, Nakagawa Y, Funada Y, Sawa S, Takeda H, Morimoto T, Koike K, Ishitani O (2014) Ring-shaped rhenium(I) multinuclear complexes: improved synthesis and photoinduced multielectron accumulation. Inorg Chem 53:7170–7180
4. Koike K, Okoshi N, Hori H, Takeuchi K, Ishitani O, Tsubaki H, Clark IP, George MW, Johnson FPA, Turner JJ (2002) Mechanism of the photochemical ligand substitution reactions of fac-$[Re(bpy)(CO)_3(PR_3)]^+$ complexes and the properties of their triplet ligand-field excited states. J Am Chem Soc 124:11448–11455
5. Rohacova J, Ishitani O (2016) Rhenium(I) trinuclear rings as highly efficient redox photosensitizers for photocatalytic CO_2 reduction. Chem Sci 7:6728–6739
6. Shvo Y, Hazum E (1975) A new method for the synthesis of organic iron carbonyl complexes. J Chem Soc Chem Commun 829–830
7. Blumer DJ, Barnett KW, Brown TL (1979) Trimethylamine N-oxide promoted reactions of manganese, molybdenum and tungsten carbonyl complexes. J Organomet Chem 173:71–76
8. Cheung AW, Lo LT, Ko CC, Yiu SM (2011) Synthesis, functionalization, characterization, and photophysical study of carbonyl-containing isocyano rhenium(I) diimine complexes. Inorg Chem 50:4798–4810
9. Morimoto T, Ito M, Koike K, Kojima T, Ozeki T, Ishitani O (2012) Dual emission from rhenium(I) complexes induced by an interligand aromatic interaction. Chem Eur J 18:3292–3304
10. Tsubaki H, Sekine A, Ohashi Y, Koike K, Takeda H, Ishitani O (2005) Control of photochemical, photophysical, electrochemical, and photocatalytic properties of rhenium(I) complexes using intramolecular weak interactions between ligands. J Am Chem Soc 127:15544–15555
11. Tsubaki H, Tohyama S, Koike K, Saitoh H, Ishitani O (2005) Effect of intramolecular π-π and CH-π interactions between ligands on structure, electrochemical and spectroscopic properties of *fac*-$[Re(bpy)(CO)_3(PR_3)]^+$ (bpy = 2,2'-bipyridine; PR_3 = trialkyl or triarylphosphines). Dalton Trans 385–395

Chapter 6
Conclusions

In this thesis, useful synthetic methods for multifarious multinuclear complexes, including the so-called "*building block approach*", which utilizes various C–C coupling reactions between metal complexes with functional groups as active spots, were reported. The results are summarized as follows:

Chapter 1. The author overviewed the conventional synthetic methods for constructing multinuclear complexes using stepwise coordination of precursor metal complexes to the multidentate bridging ligands and pointed out the synthetic problems in the ordinary synthetic methods. The new synthetic methods using coupling reactions were proposed to enlarge the variety of multinuclear complexes in this chapter.

Chapter 2. The Mizoroki–Heck reaction was first applied to the coupling of different kinds of emissive metal complexes. By using two types of metal complexes as starting materials, i.e., metal complexes with vinyl groups and those with bromo groups, various hetero-trinuclear complexes were successfully synthesized. The reactions selectively proceeded by heating a MeCN solution containing the metal complexes, $Pd(OAc)_2$, PPh_3, and AcONa. During the reaction, injection of a small amount of air was required to promote the reaction. This method could be successfully applied to the synthesis of various trinuclear complexes with strong absorptivity in a wide range of visible light and long emission lifetimes. The $Ru(II)_2$-Re(I) trinuclear complex showed photocatalytic ability for CO_2 reduction even though the selectivity and the durability of the photocatalysis were not high.

Chapter 3. The reaction conditions for the Mizoroki–Heck reactions with high selectivity and the homo-coupling reactions were discussed. The homo-coupling reactions proceeded selectively in a MeCN solution containing complexes with bromo groups, $Pd(OAc)_2$, and AcONa by heating for several days. Under the same reaction condition, the Mizoroki–Heck reaction proceeded with high selectivity in the presence of both a metal complex with bromo groups and an excess amount of metal complexes with vinyl groups. Phosphine ligands improved the selectivity of the Mizoroki–Heck reaction even though they were oxidized by injecting air after heating the reaction solution for several hours. By utilizing a combination of these

Y. Yamazaki, *Development of Synthetic Methods for Novel Photofunctional Multinuclear Complexes*, Springer Theses, https://doi.org/10.1007/978-981-16-7148-7_6

three selective coupling reaction conditions, various novel multinuclear complexes, which could not be selectively obtained by using the ordinary synthetic methods, were successfully obtained in reasonable yields.

Chapter 4. The vinylene or ethynylene linkers in the bridging ligands of the multinuclear complexes synthesized by various coupling reactions, e.g., the Mizoroki–Heck reaction, olefin metathesis, and Sonogashira coupling, could be successfully converted to the corresponding saturated carbon chains using the photochemical hydrogenation reaction. This hydrogenation improved the photocatalytic abilities of the multinuclear metal complexes for CO_2 reduction. The hydrogenation proceeded in a MeCN-pyridine-CF_3COOH (v/v/v = 3:1:0.1) mixed solution containing starting metal complexes and a sacrificial electron donor under visible light irradiation with high selectivity. The new hetero-trinuclear complex **Os-Re-Ru** was successfully synthesized by using stepwise Mizoroki–Heck reactions and the hydrogenation reaction. This complex showed strong absorptivity over a wide range of visible regions because of the three different units. The TON_{CO} of **Os-Re-Ru** was higher than those of **Os-Re** and **Ru-Re**. The hydrogenation was also proceeded by the bulk electrolysis in the presence of a proton source and a base. These hydrogenation reactions should be applied for various multinuclear complexes with multiple bonds in the bridging ligands.

Chapter 5. Ring-shaped multinuclear Re(I) complexes with a bromo group were synthesized by using thermal CO-ligand substitution reaction with Me_3NO. The Re(I) ring was successfully connected with a Ru(II) tris-diimine complex or another Re(I) ring by using the Mizoroki–Heck reaction or the homo-coupling reaction. These results clearly indicate that multinuclear complexes can also be used as building blocks for the synthesis of higher-order multinuclear complexes. The synthesized multinuclear complexes with Re(I)-ring units showed efficient light-harvesting functions.

From the results described in this thesis, various metal complexes, i.e., not only mononuclear complexes but also multinuclear complexes, will be selectively connected with other metal complexes for constructing various multinuclear complexes. By combining these coupling methods with the hydrogenation reactions, the binding mode of the linkers in multinuclear complexes can be modified for tuning the photophysical properties and photocatalytic ability. Therefore, the synthesized multinuclear complexes will be able to be used for various purposes, such as photocatalysts, photosensitizers, light-absorbers, and light-harvesting systems. The synthetic methods and strategies developed in this thesis should enlarge the variety of multinuclear complexes in terms of not only the structures but also their functions.

Curriculum Vitae

Yasuomi Yamazaki Department of Materials and Life Sciences
Faculty of Science and Technology
Seikei University
3-3-1 Kichijojikitamachi, Musashino-shi, Tokyo 180-8633, Japan
email: y-yamazaki@st.seikei.ac.jp
Web: https://researchmap.jp/y-yamazaki/

Appointments

- Postdoctoral researcher, Tokyo Institute of Technology (April 2016–March 2018)
- Assistant Professor, Seikei University (from April 2018)

Education

- Doctor of Philosophy (Ph.D.) in Science, Tokyo Institute of Technology (2013–2016), advisor Prof. Osamu Ishitani
- Master of Science (MSc) in Chemistry, Tokyo Institute of Technology (2011–2013), advisor Prof. Osamu Ishitani
- Bachelor of Science, Tokyo Institute of Technology (2007–2011)

Research Statement and Interests

My research aims to explore photofunctional transition metal complexes for photoluminescence, redox-photosensitizing reactions, and light-harvesting systems. My main contributions are on the development of various synthetic methods for new classes of photofunctional multinuclear complexes. I am also interested in developing photocatalytic systems for CO_2 reduction and investigating the reaction mechanism of CO_2 reduction catalyzed by metal complexes.

Y. Yamazaki, *Development of Synthetic Methods for Novel Photofunctional Multinuclear Complexes*, Springer Theses, https://doi.org/10.1007/978-981-16-7148-7